Improve Your Flying Skills
Tips from a Pro

Other Books in the TAB PRACTICAL FLYING SERIES

ABCs of Safe Flying—2nd Edition *by David Frazier*

The Pilot's Radio Communications Handbook—3rd Edition
by Paul E. Illman and Jay Pouzar

Aircraft Systems: Understanding Your Airplane
by David A. Lombardo

The Art of Instrument Flying *by J.R. Williams*

The Aviator's Guide to Flight Planning
by Donald J. Clausing

Mountain Flying *by Doug Geeting and Steve Woerner*

Avoiding Common Pilot Errors: An Air Traffic Controller's View
by John Stewart

The Beginner's Guide to Flight Instruction—2nd Edition
by John L. Nelson

The Pilot's Air Traffic Control Handbook
by Paul E. Illman

Ocean Flying—2nd Edition *by Louise Sacchi*

Mastering Instrument Flying
by Henry Sollman with Sherwood Harris

Improve Your Flying Skills

Tips from a Pro

Donald J. Clausing

TAB Books
Division of McGraw-Hill, Inc.
New York San Francisco Washington, D.C. Auckland Bogotá
Caracas Lisbon London Madrid Mexico City Milan
Montreal New Delhi San Juan Singapore
Sydney Tokyo Toronto

Disclaimer This publication contains chart illustrations that have been reproduced with the permission of, and are copyrighted by, Jeppesen Sanderson, Inc. They are for purposes of illustration only, and in most cases have been enlarged or reduced. They are not to be used for navigation. Chart information was current at the time of writing, but is subject to change.

In addition, this publication contains information that is supplied and copyrighted by Beech Aircraft Corporation. It is provided for educational purposes only and is not to be used under any circumstances in the operation or maintenance of an actual airplane.

© 1990 by **Donald J. Clausing**.
Published by TAB Books.
TAB Books is a division of McGraw-Hill, Inc.

Printed in the United States of America. All rights reserved. The publisher takes no responsibility for the use of any of the materials or methods described in this book, nor for the products thereof.

pbk   8 9 10 11 12 13 FGR/FGR 9 9 8
hc   1 2 3 4 5 6 7 8 9 FGR/FGR 9 9 8 7 6 5 4 3 2 1 0

Library of Congress Cataloging-in-Publication Data
Clausing, Donald J.
   Improve your flying skills : tips from a pro / by Donald J. Clausing
      p.   cm.
   ISBN 0-8306-8328-3   ISBN 0-8306-3328-6 (pbk.)
   1. Airplanes—Piloting. I. Title.
TL710.C544   1990
629.132′52—dc20   89-77182
                                                CIP

Acquisitions Editor: Jeff Worsinger
Technical Editor: Norval G. Kennedy
Director of Production: Katherine G. Brown
Book Design: Jaclyn J. Boone
Cover photograph courtesy of Piper Aircraft Corporation.

Contents

Introduction ix

1 The Basics 1
IFR Versus VFR • Straight-and-Level • Trim • Altitude Control • Turns • Climbs • Leveling-Off • Descents • Airspeed Control • Autopilots • Holding • VOR Tracking • Flight Service • Limitations • Weight and Balance • Regulations • Clearances • The System

2 Flight Planning 25
VFR and IFR Flight Planning: Differences • Major Components • Destination • Preferred Routes • Distance • Altitude Selection • Power Selection • Estimating Time and Fuel • Alternates • Reserves • Aircraft Fueling • Flight Logs • Maximum Range Trips • Weight and Balance • Flight Planning as Planning

3 Cruise Control 45
En Route Performance • En Route Options • Long-Range Cruise • Very Low-Power Cruise • Low-Power Cruise • Normal Cruise • High-Speed Cruise • Maximum Cruise • Lowest Cost Cruise • Conclusion

4 Approaches 63
Circling Approaches • Non-Precision Approaches • NDB Approaches • ILS Approaches • PAR Approaches • Safety Versus the Mission

5 Aircraft Limits 83
Contingencies and Backups • Basic Limitations • Engine Reliability • Overwater • Inhospitable Terrain • Urban Area: Takeoff and Landing • Single-Engine IFR • Single-Engine Night Flying • Multiengine Takeoff Performance • Takeoff Alternates • Engine-Out Climb Gradients • Turboprops • Jets • Weather Limitations • Common Sense

6 Personal Limits 99
Privileges and Limits • Instinct • A Systematic Approach • New Private Pilots • Instrument Training • Day Trips • Extended Cross-Country Flying • Limitations of Visual Flight Rules • Benefits of Instrument Flight Rules • New Instrument Rated Pilots • The New Captain Principle • Anxiety • Upgrading • Low Ceilings and Visibilities • Conclusion

7 The Weather 111
Theory and Practice • Time Element • Forecasting • The Briefing • FAA Standard Weather Briefing • Alternates • Telephone Briefings • Computer Briefings • Conclusion

8 Communications 127
Point of View • Limitations of Radio Communications • Data Links • Communication Problems • Clarity • Sidetones • The Language • Key Words • Verification • Frequency Management • Squelch and Volume • Memory Tricks • Simultaneous Transmissions • Radio Procedures: The Key to the System

9 Emergency and Abnormal Procedures 141
Checklist Oriented Problem Solving • Checklist Preparation • Emergency Verses Abnormal • Memory Items • Emergency Checklists • Abnormal Situation Checklists • Professionalism

10 Proficiency 161
The Process • Acquisition • Margin of Safety • Checkrides • Recurrency Training • Maintaining Proficiency • Simulator Training • Frequency • Systems • Procedures Training • Flight Training • Loss of Proficiency • Regaining Proficiency • Excuses, Egos, and Rewards

11 Judgment 173
Insight • Judgment as Learning • Flying Judgment • Gray Areas • Essential Elements • The Hard Calls • Conservatism • Flight Tests

12 Strategy 183
BILAHs • Briefing • IFR • Logs • Alternates • Hazardous Weather • Conclusion

13 Flight Levels — 199
Serial Learning • Background • Differences • The Core • Instruction • Commercial Flying • The Right Seat • The Left Seat • The Right Seat, Again • Pilot-in-Command

Index   207

Introduction

I'VE LIKED AIRPLANES FOR AS LONG AS I CAN REMEMBER. THE FIRST AIRPLANE I WAS able to see up close was an Army observation plane (an L-4, I think). I was 6 years old and my father, who was in the Army, had taken my three-year-old brother, Dean, and me out to the airfield to look at the airplanes. (We were living in Japan then.) He held us up to see inside an airplane and he explained how the stick worked, and what the pedals were for, and where the throttle was that made it go, and it all looked pretty simple.

The first airplane I ever flew on was a Boeing Stratocruiser from Seattle to Minneapolis as part of our return from Japan. In those days—1953—passengers were still allowed to visit the cockpit in-flight and talk to the pilots, which I did, of course. I remember the cockpit as a great, huge greenhouse affair with controls and indicators and switches all over the place. I was absolutely thrilled by the whole thing, and I decided then and there that being a pilot had to be the greatest thing in the world a person could possibly be.

The first thing I did at our new house in the states was to commandeer a shelf under the basement stairs and turn that space into a cockpit. My father helped me nail a couple of seats to the shelf, and I nailed some sticks together to make steering wheels, and then my brother and I used crayons to draw all the dials on the underside of the stairs and we had our own Boeing Stratocruiser.

Dean and I used to sit down there for hours, rocking the steering wheels back and forth, talking to each other out of the side of our mouths—we thought that made it sound like we were talking on the radio—and pretending we were operating the controls and switches. I was the pilot because I was the oldest. Dean was the copilot. Sometimes I let Dean fly, but he was only three, so I couldn't let him be the pilot for very long or we would have crashed. He seemed to accept that all right.

Introduction

I'd say things like, "Roger, (I thought all pilots were named Roger) check the dials, I think we have a problem with one of the motors." "Check the map, I think we're over dangerous country." When I couldn't think of anything else to say: "Pilot to copilot" and he would say "Roger, pilot to copilot" and then we'd go back to making humming sounds like motors.

Fortunately, by the time I was old enough to fly for real, airplanes were a lot easier to fly than that Boeing Stratocruiser. I still don't know how pilots learned what all those dials and switches did. I do remember, though, that there were about seven guys up there to help. It must have been quite an airplane.

Today it doesn't take nearly as many people to fly an airliner as it did then, but it still takes at least two pilots, and often three. Flying started out as a simple, single-pilot affair, of course—Wilbur and Orville had to flip for the first flight. Flying became more complex and sophisticated and more people found their way into the cockpit of larger and larger airplanes: copilots, third pilots, navigators, flight engineers, mechanics, mechanics' helpers, radio operators, radio operators' helpers. At least one airline, Pan Am with the original Clipper flying boats, even had captains, like on a ship: a senior, non-flying pilot who acted as the aircraft commander. Junior pilots did the actual flying. It didn't work very well, though, because pilots need to fly and Pan Am abandoned the practice as the flying boats were phased out.

The nautical model that served aviation so well in its early days eventually reached its limits, but we still fly airplanes with rotating beacons and position lights similar to those found on boats and ships, we still find our way with "navigation," we still land at "airports," and we still call the pilot-in-command of a transport category aircraft a "captain" because of these nautical roots.

At some point, though, increasing aircraft complexity and sophistication began to decrease the workload in the cockpit instead of increasing it. The first casualty of aeronautical automation was the radio operator. (There it is again: aero nautical.) Radio operation became so simple that the copilot was able to take on communications as an additional duty.

As cruising speeds increased and as aircraft systems and engines became more automated and reliable, relief pilots and mechanics eventually became unnecessary and then, first over land and later over water, the navigator was phased out as the captain was able to assume simplified, electronic navigational duties with the assistance of the copilot. Relatively small, two-engine jet airliners with two-man crews—the DC-9 and the Boeing 737—were eventually developed to replace the DC-7s and Constellations, or Connies, both of which required flight engineers, and then, first with the MD-80 series and later with the Boeing 757/767 and the Airbus A300/310, the flight engineer was eliminated on large twin engine jets.

Now, with the recently certified two-pilot Boeing 747-400, the two-man crew has become the standard for all transport category aircraft, regardless of aircraft size or number of engines. The flight engineer is now an endangered species, and with that the cycle in transport crewing has come almost full circle.

Introduction

But nobody is even hinting at the possibility of removing the second pilot—the copilot (or first officer as he is usually called in the airline business)—from transport category airplanes, although there is no reason to think that it couldn't theoretically be done: If a particular Cessna Citation can be flown single-pilot, then a 747 might be flown single-pilot also; the 747 cockpit would have to be rearranged so that all the essential controls, the gear and flap handles, for instance, could be reached from the left side, just like the Citation, but it could be done.

Through all of this the private pilot has been flying along by himself, and while this might seem a little inconsistent, it is one of those inconsistencies that makes sense because it has to; as a practical matter, requiring more than one pilot in general aviation aircraft would effectively legislate personal flying out of existence.

The reason it works for general aviation is simple: while a two-man crew might be desirable, two pilots are not, in fact, absolutely necessary to safely operate normal and utility category aircraft. Having only one pilot doesn't work for transport category aircraft, not so much because it can't, but because it *doesn't* have to—the airlines and corporations that operate transport category aircraft can afford two pilots and the redundancy has come to be expected by management, passengers, the FAA, and the pilots themselves, not to mention insurance companies.

When a company I once flew for added a Citation I/SP (a single-pilot Citation) to its fleet, there was initially some talk about actually operating it with a single pilot for certain easy trips in good weather. Shortly after it arrived, though, but before anything had been done about changing the normal two-pilot policy, a copilot passed out during climb out. He came around a minute or so later (after the captain had declared an emergency and initiated a return to base), but that one incident ended all further talk of single-pilot operations. No cause was ever found and the pilot never passed out again. (He was new and he probably didn't know enough not to eat at the airport.)

This example illustrates an obvious point: even if transport category aircraft are someday so simple to operate that the copilot becomes completely unnecessary, the aircraft will still be operated with two pilots, if for no other reason than to have a pilot in reserve.

Another reason transport category aircraft will always be operated with two pilots—one that no one ever seems to talk about much but one that is really much more important—is that *there is no better way to train copilots to become captains than with a two-pilot system.*

The pilot-copilot system is, in reality, an apprenticeship system. The copilot has duties, of course, and these normally include handling the radios, reading the checklists, maintaining the flight log, and managing other paperwork. But is is also customary to "swap legs," that is reverse these duties and let the copilot fly every other leg under the captain's control and supervision, so the copilot learns those duties, too. There is no better way for a new pilot to hone flying techniques, perfect procedures, and acquire good judgment, than under the watchful and critical eye of a more experienced senior pilot.

Introduction

Copilots also learn by observation. Flying isn't so much a matter of physically moving a large object through the air as it is a matter of making an endless series of decisions. In a two-pilot operation the copilot is given the opportunity to spend many hours watching another, presumably more experienced, pilot make those decisions. He is able to test himself against the captain's decisions and to listen to and learn from the captain's line of reasoning. Copilots learn even when captains make mistakes, as they inevitably do. No other system has been devised that teaches copilots to be captains as well as this one.

Private pilots don't get a chance to learn this way. Private pilots are instant captains, turned loose after as little as 35 hours of instruction and one checkride to teach themselves, mostly by trial and error, the lessons necessary to safely and routinely operate their aircraft under a variety of circumstances and conditions. They do not have a more experienced pilot to observe, and they do not have a copilot to help. It's a very tough job.

I worked for a jet charter and aircraft management company at one point in my career, and the vice president of that company, a pilot himself who used to fill in now and then as a copilot, told me a great story. He had recently convinced his wife that she ought to take flying lessons, and she had just returned home from doing some solo practice in the local area and was complaining that learning to fly was hard work. He listened, no doubt sympathetically, and explained that everyone feels that way in the beginning, but it wouldn't always seem hard, and, after all, it couldn't be all *that* hard—she *was* flying a pretty simple airplane, unlike the jets he flew. She smiled sweetly and said, "That's right. But you're just a copilot. *I'm* the pilot-in-command."

The general aviation pilot-in-command has the toughest job in aviation and very little help to make it easier. No one is ever there to help him (or her) when necessary and it shows, unfortunately, in the accident record. The reason professionals fly the way they do is because they were taught to fly that way by other pilots.

The general aviation pilot is not part of this loop. This book is an attempt to correct that. This book attempts to put the non-professional pilot in the copilot's seat for awhile—a position where he *can* learn from the experiences and mistakes of other pilots—only in this case, the experiences and mistakes are mine.

1
The Basics

AMATEUR PILOTS DO NOT EXIST. SOME PILOTS ARE PAID TO FLY AND OTHERS AREN'T, but all have to do the same job, regardless of whether or not anyone is paying them to do it. Everyone—professional and non-professional—is part of the same system and everyone has to cooperate within that system. Everyone has to meet certain minimum standards of performance for the system to function properly and safely. I call these minimum standards of performance "The Basics."

In practical terms "The Basics" are the things you must be able to do in order to operate an aircraft safely, routinely, and reliably on a daily basis. They consist of the familiar fundamentals of aircraft control:

1. straight-and-level
2. turns
3. climbs
4. descents
5. level-offs
6. altitude control
7. airspeed control

Every pilot must be able to communicate using VHF radio, he must know what a good weather briefing is and know how to get one, he must be able to flight plan properly, and he needs to know how the flight service station system works in order to be able to obtain updated weather en route. He must also be fluent with the contents of his approved aircraft flight manual, especially the aircraft limitations section, be able to verify proper

THE BASICS

weight-and-balance, know and understand all appropriate regulations, and he must know as thoroughly as possible the emergency and abnormal situation procedures for his aircraft.

Flight under instrument flight rules (IFR, as compared to flight under visual flight rule, VFR), requires that the pilot must be able to control the aircraft using only the aircraft instruments for information, he must be able to shoot both precision and non-precision instrument approaches, enter and establish the aircraft in holding patterns from any angle, and be able to navigate using VORs and the Victor airway system.

Most of these skills will be covered individually in this chapter. A few are important enough (communications, flight planning, approaches, weather, and emergency and abnormal situation procedures), to warrant separate chapters. These are the fundamental skills required of all pilots to function properly in the system—the basics.

IFR VERSUS VFR

This book is primarily concerned with flying as transportation—the safe, routine and reliable movement of people and things from point A to point B. In theory, at least, that can be done VFR as well as IFR. In practice, flying as transportation means flying IFR. Private pilots often don't like to hear or accept this, but it is still true.

The problem with trying to use VFR for routine cross-country transportation is that in order for VFR flying to be safe, the weather has to be excellent over the entire route, and the weather has to stay that way once you get there or you can't get home. This is seldom the case—not both ways anyway—and it puts tremendous pressure on the pilot trying to make it work.

Even if you live in an area that always has great weather, the advantages provided in filing IFR are numerous: continuous radar monitoring (in the U.S.), continuous availability of traffic advisories and IFR aircraft separation, automatic avoidance of restricted areas, automatic opening and closing of flight plans, easy access to TCAs, flight following, terrain monitoring, and instant assistance in the event of an in-flight emergency. The advantages make having an instrument rating and filing IFR an essential part of making flying safe, reliable, and routine.

Therefore, the first step in mastering "The Basics" is to get an instrument rating, even if you have just received your private pilot's certificate. It's fun, it's not half as hard as you think and the rating will do wonders for your basic flying skills. The sooner you start working on an instrument rating, the sooner you can start reaping the benefits.

This book does cover VFR flying because it would be deficient otherwise. One constant theme found throughout this book is that filing an instrument flight plan does not make flying harder, it makes flying easier and, done properly (which is not hard to do if currency is maintained by always filing IFR, regardless of the weather), it also makes flying safer. There will be more on this subject in later chapters—at this point all I ask is that you simply keep this thesis in mind.

STRAIGHT-AND-LEVEL

The most basic of "The Basics" is straight-and-level. Straight-and-level means being able to maintain altitude within 100 feet of the desired altitude (that's the level part) and being able to hold a heading within five degrees of the desired heading (that's the straight part). In smooth air you should be able to do better than that: 50 feet and two or three degrees is not unreasonable. Those are the standards for straight-and-level and the system is predicated on an expectation that you can meet those standards.

TRIM

The better the airplane is trimmed (meaning the more completely control pressures have been removed) the easier it is to stay within these limits of plus or minus 100 feet and five degrees. I like to use the manual trim for the elevator (FIG. 1-1) when I have a choice; some airplanes only have an electric trim. Most electric trims are just add-ons to the manual trim and they usually jump and jerk around and generally don't work very well. They're okay for big trim changes, but not for fine tuning. With the manual trim you can use little nudges to fine tune the elevator exactly where you want it. Smooth out the last that little bit of fingertip pressure on the control column so that the airplane will maintain airspeed and altitude hands off or, in rough air, return to trimmed airspeed and altitude hands off.

After the elevator trim is set for level flight work on the rudder and aileron trim. Always start with the rudder because it is virtually impossible to trim an airplane properly about all three axes of you start with the ailerons. Trim the rudder by taking your feet off the pedals and physically holding the wings level using whatever control force is necessary. Trim the ball to the center as a starting point, then note the heading and watch for any drift off of that heading. (Wait at least a minute, two or three is better. A heading bug may be positioned as a reference.)

If the heading changes, that means the airplane is turning and with the wings level, the only thing that can be causing a turn is rudder deflection; trim it out.

(Asymmetric power could also cause a turn, but it is assumed that the aircraft is either single-engine, or if multiengine, that the power has been set equally on each engine. In any case, even if the power has not been set equally, the rudder should still be trimmed to hold whatever rudder deflection is required by the asymmetric power.)

When the rudder is trimmed and any drifting tendency has been removed, trim out any aileron control force you've been holding to keep the wings level. Finally, go back and recheck the elevator trim. Correcting the rudder and aileron trim will result in a cleaner aircraft that will probably necessitate an elevator trim change.

ALTITUDE CONTROL

Once the airplane is properly trimmed about all three axes expect very little trouble staying within plus or minus 100 feet of altitude, even in fairy rough air. Very few pilots

THE BASICS

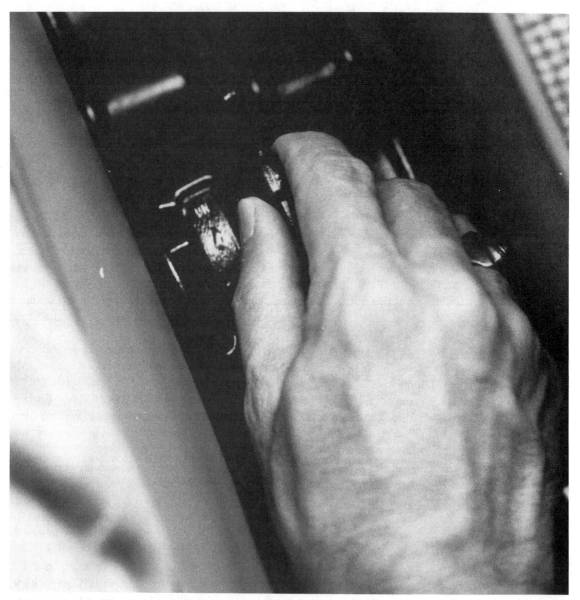

Fig. 1-1. *Fine tuning the elevator trim can usually be done best with the manual trim; electric trim, when available, is best for major trim changes, not for fine tuning.*

realize how critical this 100 feet of leeway is. This is the maximum deviation the FAA can allow and still keep VFR and IFR airplanes separated. (IFR aircraft fly at 1,000-foot intervals and VFR aircraft fly 500 feet above or below the IFR intervals.) Quick arithmetic will show why.

Altitude Control

Suppose that a VFR aircraft with a desired cruising altitude of 9,500 feet is actually flying at 9,600 feet, 100 feet high, which is the maximum legal allowable deviation (FIG. 1-2). Also suppose an IFR aircraft with an assigned altitude of 10,000 feet is actually flying at 9,900 feet, 100 feet low, again the maximum allowable legal deviation.

So far so good. Three hundred feet of separation still remains between the two aircraft at 9,600 feet and 9,900 feet, respectively. However, the ATC system also allows for 75-foot altimeter error because the altimeter is a simple pressure measuring device and much greater accuracy than that is not practical. If both aircraft altimeters are off by 75 feet in the wrong direction—that is, if the VFR aircraft is actually at 9,675 feet (although indicating 9,600) and the IFR aircraft is actually at 9,825 (although indicating 9,900)—they will pass within 150 feet of each other (9,825 minus 9,675) and be perfectly legal.

Every time you deviate from the proper altitude by more than 100 feet you reduce that separation even more. If each pilot is off by more than an additional 50 feet, for instance 150 feet off of the cruising altitudes, then the separation would be further reduced to 50 feet, which is close enough for a heart attack. The errors might partly or completely cancel, or they might go in the right direction rather than the wrong direction. Playing the odds is not supposed to be part of being a good pilot. Think about that every time your altitude slips just a little above or below the allowable limit.

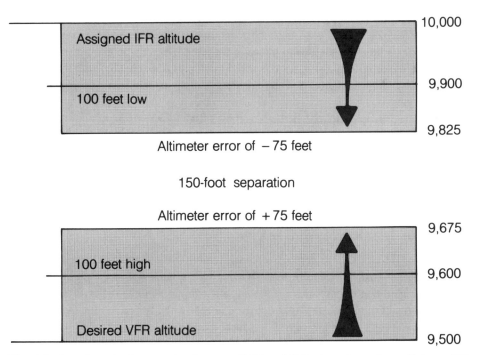

Fig. 1-2. *Legal vertical separation between VFR and IFR aircraft can be as little as 150 feet.*

TURNS

Turns are very important to smooth, safe flight. In fact, *smooth* flying and *safe* flying are generally the same thing. Pilots flying VFR usually don't have problems with turns, other than letting the altitude slip a little, but non-instrument rated pilots often do get in trouble when turning in clouds to find VFR conditions. The reason for this is that turns have to be precise, with frequent cross-checks of the airspeed and altitude. They are not at all hard to do, but they do take a little training and practice, and this is yet another good reason private pilots should take instrument training.

For instrument rated pilots, using the proper amount of bank as appropriate to the situation is the key not only to maintaining aircraft control, but also to fitting in with the instrument system: flying vectors, tracking airways, shooting approaches, and entering holds. Let's take a look at some of those situations.

One or Two Degrees of Bank

The smallest amount of bank possible, one or two degrees, should generally be reserved for shooting approaches and for VOR tracking. When possible, watch a Flight Director work because it is a computer that determines the proper bank angle for a given situation and displays that angle with command bars on the pilot's attitude indicator. When established on an inbound localizer course, and bracketing a heading within five degrees, all bank commands will be very small but continuous: the command bars will be constantly moving a little bit at a time to maintain the aircraft exactly on the inbound course. The flight director will almost never command the wings to be exactly level, but neither will it command large corrections, again assuming that the inbound course has been properly intercepted.

The same principle applies whether you have a flight director or not: once established on an inbound localizer approach course, use continuous bank corrections of one or two degrees in order to adjust the heading by one or two degrees and continue exactly on the centerline of the approach. These same bank angles will also work well for final course corrections on VOR approaches and for fine tuning of headings when tracking between VORs.

Five to Ten Degrees of Bank

En route, at altitude, 10 degrees of bank should be the most ever used for routine turns to change heading or course, and five degrees of bank will normally be enough. Most passengers don't appreciate a big bank after sitting straight-and-level for awhile. Passengers might think that airplanes are like canoes; when tipped too far they will flip upside-down. That's not true of course, but there's no need for a big bank just to change a heading a few degrees; five or 10 degrees is plenty.

Fifteen Degrees of Bank

Fifteen degrees of bank for shallow turns should be used at minimum airspeed, such as shortly after takeoff. The difference between the stalling speed straight-and-level and the stalling speed at 15 degrees is negligible. Fifteen degrees of bank is also a good angle for 20- to 30-degree course changes en route. Fifteen degrees of bank is not so steep that it concerns the passengers, yet it results in a fast turning rate to establish a new heading quickly.

Twenty-Five Degrees of Bank

On climb-out and on descent and for major course changes en route, 25 degrees of bank is about right. Twenty-five degrees of bank will be very close to a standard-rate turn (one degree per second) at most speeds, and it is easier to fly a constant angle of bank than it is to keep referring to a separate instrument in order to fly a perfect standard-rate turn.

Twenty-five degrees is typical of the bank angle programmed into most flight directors, which is a good endorsement, and it seems to be approximately what the controllers expect in these circumstances. That's important, because the controller assumes you will turn in a predictable manner and plans all traffic accordingly. A very tight turn, at more than 30 degrees of bank, or a wide turn at seven or eight degrees of bank, might force a controller to adjust traffic.

Thirty Degrees of Bank

Thirty degrees of bank is the steepest turn normally made on instruments. It's a good rate of turn without getting into the control problems of steep turns (45 degrees or more). A thirty-degree-bank turn is especially handy when turning to final on a circle-to-land approach. The key to a good circling approach is to stay in close but not over-shoot the final. (See Chapter 4.) Doing this usually requires tight turns. In fact, if your initial turns are done at 25 degrees, a 30-degree turn to final will generally ensure staying inside the final approach course and it is much easier to intercept the final approach course from the inside than to overshoot and try to recapture it. Thirty degrees of bank is also about the correct amount of bank to use when a controller asks for a "good rate" to a new heading. There is no need to overreact to a request for a good rate because a good standard rate turn will usually satisfy the request, but 30 degrees ensures a good rate without overdoing it and risking control problems.

Forty Five Degrees or More of Bank

A good steep turn—45 to 60 degrees of bank, altitude within 100 feet, and airspeed within five knots—is one sign of a good pilot, which doesn't make you a bad pilot if you have trouble with it because we all do from time to time. Practice shall suffice. Steep turns are excellent exercises because they demand concentration, precision, and control, and they are not very forgiving, which doesn't mean that they are dangerous, it just means that

THE BASICS

small errors will be readily apparent. Beyond exercises, there are times when a steep turn is necessary: conflicting traffic, terrain avoidance, severe weather avoidance, or a controller request for an "immediate" turn. Being able to do a good steep turn is therefore more than just an exercise, but is one of "The Basics."

The most common mistake pilots make while performing steep turns is to add back pressure too soon. You really don't need any back pressure until going through approximately 30 degrees of bank. Whether it is because rolling into the bank generates a small amount of lift, or whether it's just a function of inertia, you don't need back pressure initially. If added too soon the aircraft will climb.

The next mistake is to focus primarily on the altitude. This almost always leads to "porpoising," or chasing the altitude, and it's awfully hard to get things back under control. The key to maintaining altitude is not to stare at the altimeter, it is to maintain the correct *pitch* attitude. Experiment to discover the pitch attitude that results in zero altitude gain or loss; for many airplanes this will be about five degrees nose up, but that is just a rough guide. In any case, it will be two or three degrees higher than level flight for that airspeed. If this sounds like nit-picking little stuff, so be it because that's what is takes to maintain altitude in a steep turn.

Back pressure is the inevitable by-product of attitude control, but it is not an end in itself; it you are holding the proper pitch attitude, you will find that you are holding some back pressure (assuming the aircraft was trimmed for straight-and-level flight before you began the turn). Speed is controlled by power; you probably do want to "bump" the power up a notch as you roll into the turn. If the bank and power are constant and if the proper pitch attitude is held—big "ifs"—the airplane will truck around the turn as if the altimeter needle was glued in place.

I used to think steep turns were dumb, but that was because I had trouble with them. I don't think they're dumb anymore, although I still have to work pretty hard at them, which means I do an awful lot of rapid cross-checking: attitude to airspeed to attitude to altitude to heading to attitude and on and on.

When a steep turn is necessary, it must be done correctly. You don't want to let a steep turn deteriorate into a spiral (not enough back pressure initially), but neither do you want to "load it up" with back pressure in a futile attempt to gain altitude (only making the bank steeper instead), nor do you want to zoom up, possibly into the clouds (too much initial back pressure), and you certainly do not want to become disoriented and progress into what the FAA diplomatically calls an "unusual attitude." The only way to be sure you can do a good steep turn when the pressure is on is to practice, at safe altitudes, when the pressure is not on.

CLIMBS

Climbs are fairly simple and most pilots do them well, but a couple of points are worth mentioning. For the first 400 feet the primary consideration in *any* airplane is to gain altitude quickly. For Normal and Utility Category aircraft, this means holding best rate-of-climb speed (V_y) as accurately as possible because that is the speed that most efficiently translates power into altitude. Until you have at least 400 feet of altitude (or whatever it takes to safely maneuver on one engine or to glide to an emergency landing site) you can't afford the luxury of a higher climb speed.

Carrying extra speed at this point might make you feel better, especially in a multiengine aircraft, but it won't translate into altitude if an engine quits. The extra speed will dissipate almost simultaneously with no appreciable altitude gain because the drag of the windmilling propeller will be greater at the higher speed. What you want immediately after takeoff, in any airplane, is altitude, and the way to get altitude is to climb at V_y.

This segment from lift-off to 400 feet is especially critical for multiengine, propeller-driven airplanes. To suddenly lose power on one engine at slow speed while close to the ground is an extremely marginal situation for a multiengine airplane.

A multiengine airplane is similar to a rowboat; lose an oar, and a rowboat will go in circles. The same thing will happen to a multiengine airplane when one engine quits, unless something is done about it. Spiralling tendencies depend on placement of the engines: the farther out on the wing or the farther away from the fuselage the engine is installed and the more power that engine is developing at the time, the greater the turning tendency will be.

An airplane has an advantage over rowboats because airplanes have rudders. The rudder can, to a certain extent, counteract the tendency of the multiengine airplane to fly in a circle on one engine. The ability of the rudder to do so is directly proportional to the speed of the airplane, because the faster the airplane is going the more air flows over the rudder and the more "power" it has to do the job of straightening the airplane out.

Conversely, a slower airspeed means the less "power" for the rudder to counteract the turning force. As the airplane gets slower and slower it eventually reaches a point where it doesn't have enough rudder power to keep the airplane from turning and at this point the airplane becomes uncontrollable.

The speed at which control is lost with full takeoff power being developed on one side of the airplane and zero power on the other is called "minimum controllable airspeed" and is abbreviated V_{mc}.

Engine Out

If an engine quits abruptly when the speed is below V_{mc}, the airplane will both turn and roll into the "dead" engine. If it happens close enough to the ground that a recovery cannot be accomplished, the consequences are invariably disastrous. It generally takes several hundred feet to recover from an abrupt loss of power on one side when the speed is below V_{mc} so the importance of maintaining adequate flying speed cannot be emphasized too much.

The problem you have in a multiengine airplane on takeoff is that you have no choice during the first 400 feet but to use maximum power and to fly slowly (V_y) to gain minimum maneuvering altitude as quickly as possible. That is exactly the combination of high power and low airspeed that makes aircraft control a problem when an engine does quit. This sounds like one of those situations where you can't win, but it's not quite as bad as it might seem: V_y, while slow, is still well above V_{mc}. (The FAA won't certify the airplane

THE BASICS

if not). If you don't go any *slower* than V_y you won't have to worry about dropping below V_{mc}, and if you don't go any *faster* than V_y you won't be giving away any altitude you might need later, which is why flying as close as you can to V_y for at least the first 400 feet is so important.

If you do not lose an engine on takeoff and find that even at the best single-engine rate-of-climb speed (V_{yse}) you are unable to climb, you might have to settle for mushing into the ground, but the chances of surviving this kind of crash are much, much better than the chances of surviving an uncontrolled, spiraling descent into the ground, and that is, unfortunately, nearly always the outcome when the airspeed is allowed to deteriorate below V_{mc} and an engine suddenly quits.

One of the many nice things about jets with engines in the rear is that their V_{mc}s are very low, which allows for a huge margin between V_{mc} and the slowest normal operating speeds: the engines are so close to the fuselage that there is little turning tendency when one quits. If an airplane with wing-mounted engines is like a rowboat, then an airplane with engines mounted close to the fuselage is like a twin screw inboard boat, and a single-engine airplane is like a simple outboard, it either goes straight ahead or it coasts to a stop.

(The Cessna Skymaster, with one engine in the front and another in the rear, both directly on the centerline, combined the best of both worlds: multiengine redundancy, with single-engine controllability. Unfortunately, the Skymaster had a lot of problems with engine cooling, which was too bad, because it was a great idea otherwise.)

All of this has to do with the first 400 feet, when the first priority is to gain altitude as quickly as possible for maneuvering in case of an engine failure: either failure of the only engine or failure of one of two engines. Once you have some altitude and all the obstacles are clear, the most important consideration is visibility; push the nose over where you can see something and also improve engine cooling.

The efficiency penalty in flying faster than best rate-of-climb is fairly small and the benefits above 400 AGL (above ground level) are fairly large. If you are planning to climb to within a few thousand feet of the service ceiling of an aircraft, you might have to drop back to V_y to keep climbing, but for most altitudes that won't be necessary. Below 10,000 feet, in visual conditions, maintaining a specific cruise climb airspeed is not so important as being able to see over the nose. Above 10,000 feet, or in the clouds, use the manufacturer's recommended cruise climb airspeed.

LEVELING OFF

A nice, smooth level-off not only is easier on the passengers, it is also essential to good altitude control, VFR or IFR. The key to good level off is to start early, trimming all the way. Exactly what "early" means depends upon the rate of climb. A common rule of thumb is to start leveling off when the number of feet remaining corresponds to half the rate of climb: if climbing at 1,000 fpm, start to level off 500 feet early. I don't think it is necessary to memorize and observe this rule exactly, but it does provide a rough idea of what "early" means. If you use common sense and anticipate the level off, you shouldn't have any problem.

Leave the power alone and let the airplane accelerate as you reach altitude. Gradually push it over and work on leveling off and getting trimmed. Once it has accelerated, then bring the power back. Ideally the passengers won't even know when you have leveled off until they hear the power reduction. This is the smooth, accurate way, it's easy on the engine, and it keeps ATC happy. Keep your finger by the trim wheel and just keep nudging it along.

I don't recommend using the autopilot to level off. There are very few general aviation autopilots with the power and smoothness to level off properly. The very newest and best systems for jet equipment will, but even many transport aircraft autopilots don't do a very good job of leveling off. If your autopilot does a decent job of it, fine, but disengage the autopilot after it has done the leveling, make sure all three axes are still in trim, and then put it back on again if you want. An autopilot that fights an out-of-trim aircraft might require early and unnecessary maintenance.

DESCENTS

The subject of descents could fill an entire chapter, but the basics are fairly simple. The first priority is a descent that works out right; i.e., one that gets you down in time, but not too early. The second priority is a descent that doesn't abuse the engine, and the third priority is a descent that is comfortable.

Most pilots use some kind of rule of thumb to plan their descents. In pressurized aircraft a common rule is to start down when the mileage is three times the altitude (in thousands). The mileage to altitude relationship is then monitored to maintain that three-to-one ratio all the way down. Thus, from 33,000 you would start down 99 nautical miles out (plus maybe five or 10 miles if you also need to decelerate) and come down fast enough to keep that ratio constant all the way down.

Unfortunately this rule won't work for unpressurized airplanes because the descent rates are too high: 1,000 to 3,000 fpm, depending on airspeed and tailwind component. A comfortable, unpressurized rate of descent is 300 fpm. This means that from 9,000 feet 30 minutes are necessary to descend to a sea level airport. (You need to descend 9,000 feet, and you come down 300 feet for each minute: 300 into 9,000 is 30). This 300 fpm rate is probably slower than you are used to using, but it is what the airlines often used back in the old, unpressurized days, and it doesn't shock-cool the engine either. The little bit of true airspeed or tailwind given up by starting down earlier than usual probably is not worth worrying about in some cases, and the advantages are considerable.

Attain maximum fuel efficiency in the descent by pulling the power back. Let the nose drop naturally, descending at cruise airspeed. As long as fuel isn't a problem, push the nose over a little and retrim for the faster airspeed to regain in the descent some of the ground speed lost in the initial climb. Just don't overdo it; most passengers are very uncomfortable with anything that looks like a "dive." Avoid the yellow arc on the airspeed gauge because smooth air has a way of suddenly becoming rough air and the yellow arc is for smooth air only.

The Basics

AIRSPEED CONTROL

In order to have complete control over the airplane, you must be able to maintain any assigned airspeed that is safe and within the limitations of the aircraft, and you must be able to do it for all configurations: clean, partial flaps, gear down, full flaps. This applies to both straight-and-level flight, and while turning, climbing, and descending.

The starting point is clean, level flight. If you know ahead of time the amount of power necessary to maintain a given airspeed, airspeed control is a very simple matter: simply set the power to the appropriate setting and wait for the airspeed to settle. A minor adjustment should be all that is necessary after that. If you can develop a rule of thumb for a particular airplane, so much the better because you won't have to memorize a bunch of power settings and airspeeds. A rule of thumb for Falcon 20s (as an example) is: total fuel flow "equals" airspeed; i.e., 2,000 pounds per hour (pph) fuel flow will yield 200 knots airspeed, clean. (Clean meaning that no landing gear, flaps, speed brakes, or drag inducing device is extended.) In the L-1011, 5,000 pph per engine will result in an airspeed of about 250 knots clean; adjusting fuel flow above and below that amount will result in airspeeds correspondingly faster and slower than that base airspeed. In a single-engine airplane a useful rule might be something like "25 and 25 [25 inches of manifold pressure and 2,500 rpm] equals 150 knots and an inch either way equals 10 knots." (I just made that up; I don't know if it works for any actual airplanes or not, but it is going to be close and it is the *kind* of rule of thumb that can help maintain assigned airspeeds.)

As you add flaps and drop the gear, additional power will have to be added to maintain a given speed. You should have a good idea ahead of time regarding how much power is needed so the airspeed doesn't vary all over the place as the gear and flaps go out. You will have to experiment, but it's good practice.

In the climb, pitch controls airspeed. If you need a faster climb speed, drop the nose, and if you need a slower climb speed, raise the nose. Retrim in each case.

In the descent, power is available to control airspeed: add power to go faster, reduce power to go slower, and retrim. If you do not retrim, the airplane will attempt to return to the original trimmed airspeed, but at a lot slower descent rate if the power was added or at a greater descent rate if it was subtracted.

Airspeed control is a primary method for controllers to separate and sequence aircraft. If you want to operate in the system you must be able to maintain any assigned airspeed within the airplane's limits.

AUTOPILOTS

If an autopilot is necessary to do any of these things, more practice is necessary. This is not to say that an autopilot doesn't have its place, it's just that you must never allow yourself to become dependent upon it. If you need an autopilot to be able to do a legitimate job of flying the airplane, then the most competent pilot in the airplane is the autopilot and the primary source of control is something other than a human being, and that's not good.

Autopilots are invaluable aids to good instrument procedures, though, and they do have legitimate uses. One such use is to maintain altitude and heading while at cruise. Another is to shoot precision approaches (ILSs) to minimums. Approaches will be covered separately, but in terms of autopilot use, when the weather goes right to the ground, it is generally better to let the autopilot shoot the approach while you monitor its progress—hands on the wheel and throttle ready to take over at the first sign of a malfunction—than it is to try to shoot the approach yourself.

(This does not, however, relieve you of the responsibilty to be able to shoot an ILS approach to minimums if you have to, which is why routine ILSs should be hand flown to maintain proficiency, but if you don't have to, it is usually better to let the autopilot shoot an approach to minimums.)

Still another legitimate use for the autopilot is to maintain aircraft control and preserve structural integrity in severe turbulence. The autopilot will do what it must to keep the airplane upright and will attempt to maintain airspeed and altitude the best it can, but autopilots are deliberately limited (according to certification requirements) to not have enough power to overstress the airplane. In other words, the autopilot won't try so hard to maintain airspeed and altitude that the resulting control forces damage the aircraft. (Damage might result anyway, unless the aircraft is slowed to maneuvering speed, but it won't be damage as a result of control inputs.) Finally, autopilots are very useful when you need to check a chart, read a checklist, or make a log entry.

HOLDING

Despite four different maximum holding speeds that depend on aircraft type and altitude, two different leg times that depend on altitude, and the fact that the FAA expects you to correct for the effects of wind on leg length and track, *holding is easy*. But entering holds is hard (FIG. 1-3, BOTTOM).

For me the key to entering a hold is orientation: knowing exactly where I am, where the hold is, and what direction I am going to go around. Once I know all that, I find it is then very easy to figure how to get myself from where I am to where I want to be.

It often helps to actually draw the hold on the chart with a pencil. I have seen professionals do this, and nobody laughed. (But they sure snicker when someone screws up and can't get into the hold properly, and if you really mess up in a sticky situation you will probably get a violation.) If you know where you are on the chart, and the hold is drawn correctly on the chart, the preferred entry will be fairly obvious.

Sometimes all you have to do is "draw" the hold on the chart with your finger, just to make sure you have it right, but in one way or another I think it is absolutely essential that you visualize the hold and your position in relation to it in order to enter it properly.

In any case, don't get rattled, take your time (slow the airplane down to allow more time if necessary because there's no point in going fast merely to hold anyway), and when in doubt, at least stay inside the holding area. Remember that the holding pattern is a large, protected area that the controller expects you to stay in. A holding clearance is, in

THE BASICS

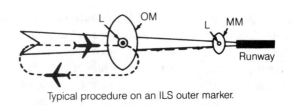

Typical procedure on an ILS outer marker.

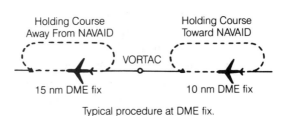

Typical procedure at intersection of VOR radials.

Fig. 1-3. *Recommended holding pattern procedures and terminology are described and illustrated in the* Airman's Information Manual, *part of which is reproduced here.*

Typical procedure at DME fix.

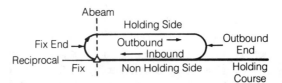

effect, an area restriction. The controller is telling you to go to your room and stay there until he calls you. The whole point of the various entries is to keep an airplane inside that area while in the process of getting established in that area. Keep that in mind and the entry should be obvious.

The *Airman's Information Manual* (AIM) says that pilots should report the time and altitude entering the hold. When the FAA says "should" they might as well say "must," so don't forget to report entering the hold. Sometimes controllers will want to terminate the radar service when established in the hold, so the report serves as a cue for them to terminate the radar service. This isn't anything to worry about—they still see you—it just makes their job a little easier. When cleared out of the hold, a controller will invariably ask for an ident and reestablish formal radar contact.

Comments

This business of holds is about 500 times more complicated than necessary, but I doubt if I'm going to get anywhere changing it all by myself. Holding procedures only manage to raise the stress level on anyone taking an instrument checkride because holding provides numerous ways to bust a ride on technicalities or unnecessary confusion and disorientation. All restrictions and requirements could be eliminated by simply expanding the holding areas so the speed and leg restrictions and the elaborate entry requirements would no longer be needed.

All entries could be reduced to either direct or parallel—direct if behind the fix, parallel if before—when the holding entry area was expanded a small amount.

Every approach should have a missed approach fix directly on the extension of the final approach course with a direct entry authorized. Unfortunately, expanding the size of all patterns and creating direct holds off every approach would be a big project, so it isn't very likely that this is going to happen.

If holding was more common I'm sure something would be done, but it's not, so I'm equally sure we're all just going to have to put up with the current holding requirements and restrictions, probably forever, so make sure you know them and review them from time to time (AIM, para 347.)

VOR TRACKING

Navigation means finding your way. For most pilots in the United States, whether, VFR or IFR, that means tracking along VOR airways (FIG. 1-4.) This is not the only way to navigate. Loran-C, in particular, is becoming a very popular non-VOR way to navigate, and NAVSTAR/GPS is not far off, but being able to navigate using VORs is still one of the "The Basics" for all pilots.

With practice, tracking along a given VOR radial can be as unconscious and habitual as driving a car down the highway, but it doesn't happen automatically. In the beginning you have to work at it. The key, as with so many things in aviation, is to make many small corrections rather than wait for a large needle deviation and have to make one or more big corrections.

It is actually much easier, and it is certainly much more accurate and efficient, to make many small corrections than a few big ones, although it might not seem easier at first and the effort might not seem to offset the increase in accuracy. But it is.

Making many small corrections doesn't mean chasing the needle, but it does mean making a small correction just as soon as the needle moves off center. Then another one. Then another one. After approximately three corrections you should have a pretty good

THE BASICS

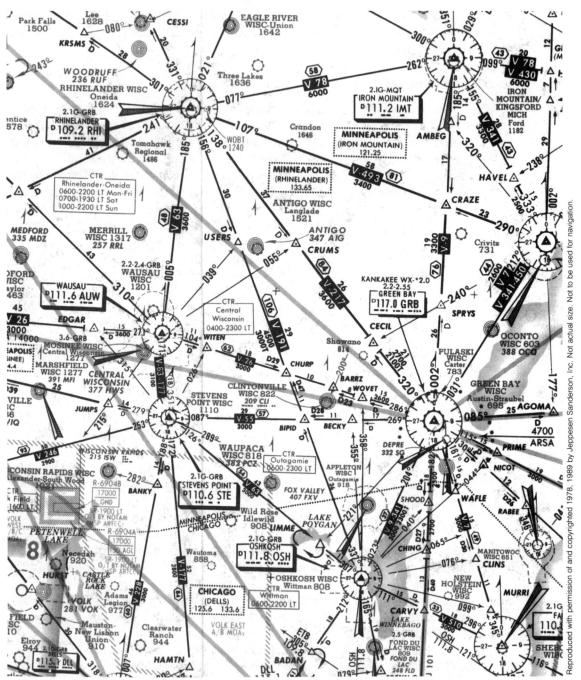

Fig. 1-4. *Instrument en route charts make VOR airway navigation and flight planning easy by providing accurate distances between fixes, radial information, and minimum safe altitudes.*

Flight Service

heading nailed down, and after that only an occasional small correction should be necessary. If the deviation needle is not allowed to get past the first dot on the VOR deviation indicator, you are on the way to accurate and easy VOR tracking.

Autopilot Assistance

When using the autopilot en route, select the "heading" mode for VOR tracking, not the "nav" mode. The "nav" mode has no brain; it will turn toward the needle until the needle starts back, at which point the needle will swing straight over to the other side, the airplane will rack itself back over the other way, the needle will swing back again, and this process will be repeated several times until the radial is finally more or less nailed down and captured. Even then it will still try to follow every subsequent swing and blip of the needle, the airplane constantly rolling back and forth chasing a wandering radial.

You can do a much better job because you are smarter. It does take a little experience to get a feel for what to ignore and what to correct, but in the end you should get a more comfortable ride than with the "nav" mode. One of the many nice things about most area navigation (RNAV) systems—LORAN-C, VOR/DME based RNAV, OMEGA, inertial nav—is "brains" to look ahead and determine what kind of correction to make and they do provide steady signals. These systems *can* use the "nav" mode if the autopilot can be coupled.

FLIGHT SERVICE

Contacting a Flight Service Station (FSS) en route is an important skill. Flight service is the best place to get up-to-date weather information en route. Many pilots also assume that FSS is the *only* place to get weather en route, but that's not entirely true: control towers can be contacted directly when within range and if they are not too busy with traffic they will be glad to provide current weather; controllers can also obtain weather, if they are not busy with other traffic. The *best* place for weather is still an FSS.

To use flight service effectively you have to know how it works and which frequencies to use for a particular service. Almost all flight service stations have a common transmit and receive frequency of 122.2 MHz. This is both good and bad. The good part is that one frequency is common to almost all FSSs and it is easy to remember: 122.2, basically, all twos. When aware of the name of a nearby FSS merely call it on 122.2. (The common call sign for all flight service stations is "radio," as in "Bridgeport Radio" for the Bridgeport FSS.) The bad part is that everybody else knows about this frequency and can remember 122.2 also, so it is usually terribly congested, often with interference from overlapping stations.

It is much better to use one of the frequencies specifically assigned to the FSS desired. These will be frequencies in the 122 range, such as 122.3, 122.35, 122.4, and the like, with one exception. These specific frequencies are found on both sectional and low altitude en route charts (FIG. 1-5). It takes a second to look them up but it generally pays off because you won't have to compete with a bunch of other pilots for air time.

THE BASICS

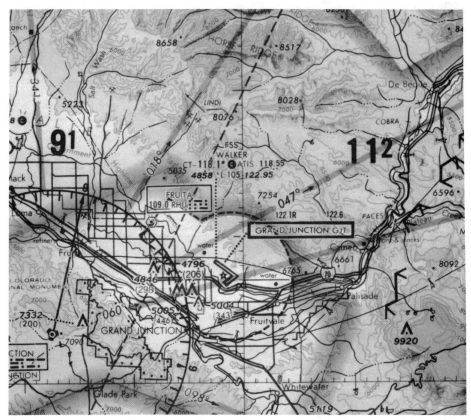

Fig. 1-5. *Specific FSS frequencies, in this case 122.6 and 122.1R (receive only) for Grand Junction, can be found on sectional charts over the heavy-line FSS identification box.*

When you do call, speak clearly in a normal conversational tone: "Grand Junction Radio, Range Rocket 1234 Xray on 122.6." With any luck Grand Junction Radio will answer. Sometimes you will get a flight service station specialist from another station with the same frequency answering and cutting out the one you want. Usually "Albuquerque Radio, disregard. 32 Xray calling Grand Junction Radio on 122.3" will do the trick.

If you do go ahead and talk to Albuquerque, Grand Junction might wake up and block Albuquerque out. (I'm not trying to pick on these two FSSs—I just pulled names off a chart.) This is not a major problem now with the new consolidated FSSs because there are fewer facilities to overlap.

FSS can provide general services: flight plan filing, complete weather briefings, NOTAMs, customs notification, and so on. If all you want is the current weather for your destination and maybe for one or two alternates, call flight watch on the common frequency of 122.0. Flight Watch has only one function, and that is to provide en route weather updates, which means you don't have to wait 10 minutes for a pilot to air file three flight plans before the frequency becomes available.

Limitations

The only FSS frequency that is not a 122. − frequency is 123.6, which is reserved for Airport Advisory Service (AAS). Airports that have an FSS, but not an operating control tower, have an AAS. This means that anyone who wants to can report his position from the airport and his intentions to the FSS, and the FSS will log all reports and keep everyone in contact informed of that reported traffic. FSS doesn't control, they just keep track of and relay the reported traffic.

In order to keep a clear channel for information that is useless if it isn't passed on quickly, 123.6 is restricted to airport advisories and you shouldn't use it to ask for the weather or file a flight plan. If you do, you will probably be politely told to change to another FSS frequency.

LIMITATIONS

Check rides for type ratings are always preceded by an oral exam. These oral exams typically focus on the Limitations section of the Aircraft Flight Manual (AFM). They might include questions on aircraft systems and performance and the examiner might include a weight-and-balance problem, but the main part of the oral exam for a type rating is usually a line by line quiz of the Limitations section of the AFM:

- This aircraft is approved for what types of operations?
- What is the minimum flight crew?
- What is the maximum number of passengers?
- What is the altitude limit for takeoff and landing?
- What is the maximum landing weight?
- What is the maximum operating speed?
- What are the limits for takeoff power?
- For maximum continuous power?

The reason you must know these things, and the reason the FAA puts major emphasis on them during a type rating checkride, is because operation within the limitations, as described in the AFM and supplemented by appropriate placards, is mandatory—it has the force of law. The limitations are not guidelines, or "good ideas," or recommendations, and they cannot be waived, except in an emergency, and even then you must be able to justify the action. Failure to observe the aircraft limitations as specified in that section of the operating manual can lead to fines, suspension of privileges, and insurance cancellation.

Obviously, in order to observe the limitations, pilots must know the limitations, and this applies to *all* pilots, regardless of whether or not they have to take a checkride to fly a particular airplane. The only way to know for sure what the limitations are is to get the AFM out, flip to the limitations section, and read it, memorize it, and then quiz yourself until you know it cold (FIG. 1-6). (Whenever I have to take an oral exam, I have my wife quiz me first. I usually know the material pretty well before I give her a shot at me, and the FAA routine is easy after that.)

THE BASICS

> **Section II** **BEECHCRAFT Bonanza F33A**
> **Limitations** **CE-674 and after**
>
> **POWER PLANT INSTRUMENT MARKINGS**
>
> **OIL TEMPERATURE**
> Caution (Yellow Radial) .... 38°C
> Operating Range
>   (Green Arc) .... 38° to 116°C
> Maximum (Red Radial) .... 116°C
>
> **OIL PRESSURE**
> Minimum Pressure (Red Radial) .... 30 psi
> Operating Range (Green Arc) .... 30 to 60 psi
> Maximum Pressure (Red Radial) .... 100 psi
>
> **TACHOMETER**
> Operating Range (Green Arc)
>   (Serials CE-674 thru CE-890 with 2- or 3-Blade
>   Propeller Installed, and CE-891 and after with McCauley
>   3-Blade Propeller Installed)
>   (Serials CJ-129 thru CJ-155) .... 1800 to 2700 rpm
> Operating Range (Green Arc)
>   (Serials CE-891 and after with 2-Blade Propeller
>   Installed)
>   (Serials CJ-156 and after) .... 1800 to 2550 rpm
> Maximum rpm (Red Radial) .... 2700 rpm
>
> **CYLINDER HEAD TEMPERATURE**
> Operating Range
>   (Green Arc) .... 93° to 238°C
> Maximum Temperature (Red Radial) .... 238°C
>
> **MANIFOLD PRESSURE**
> Operating Range (Green Arc) .... 15 to 29.6 in. Hg
> Maximum (Red Radial) .... 29.6 in. Hg
>
> **FUEL FLOW**
> Serials CE-674 thru CE-928; CJ-129 thru CJ-155:
> Minimum (Red Radial) .... 1.5 psi

Beech Aircraft Corporation.

Fig. 1-6. *The limitations section of the Aircraft Flight Manual, a part of which is shown here for the Beech Bonanza F33A, describes mandatory operating limitations that cannot be waived or exceeded except in an emergency. For Educational Purposes Only. Not to be used under any circumstances in the operation or maintenance of an actual airplane.*

It is important to differentiate between what the manufacturer *recommends,* and what the FAA *requires.* The manufacturer provides pages of information on how to set the power for various cruise conditions, how to lean the engine, when to turn fuel pumps on and off, guidelines to use for setting the cowl flaps, and so on. These are the manufacturer's recommendations and to the extent they were developed by a small army of engineers and substantiated by thousands of hours of operational experience, they are to be taken very seriously. But they are not mandatory.

You do not *have* to set the power according to any cruise charts, unless those charts are found in the AFM. If the only power limitation in the manual, on markings, or on placards is "Max power, all altitudes: 2,700 rpm," you may legally run the engine at 2,700 rpm until the fuel runs out, even if that means 95 percent or 100 percent power.

This is an important distinction, because it is very important to know where you have flexibility as a pilot, and where you do not. Limitations are mandatory, recommendations are not. The only exception to the absoluteness of the limitations is emergency authority as pilot in command to deviate from any regulation (and therefore any limitation) if necessary; but the burden of proof regarding the wisdom of that deviation will be on the pilot. Recommendations are just that—recommendations—and if you want to deviate from them that is your business and you don't have to justify those deviations to anybody except yourself, but don't do it on whim.

There is an implied requirement that the necessary equipment to observe the limitations be installed and functioning. Thus the flight manual might not specifically state that an aircraft must have a fuel pressure gauge, but if the flight manual has a maximum or minimum fuel pressure limitation, the aircraft must, by implication, have a working fuel pressure indicator because there is no other way to observe it. Likewise, if the cylinder head temperature is limited, then an aircraft must have a cylinder head temperature gauge. However, if the temperature is not limited, a gauge is not necessary and if you have one anyway and it is not working, you can still legally operate the aircraft even though the manufacturer might *recommend* that cylinder head temperature be limited to a certain maximum. (You might not want to anyway, but you can.)

The observance of the limitations, as listed in the aircraft flight manual, is one of "The Basics." The limitations define the safe, normal operating envelope of the aircraft. Any operation outside of that envelope is uncharted territory, the realm of the test pilot. Good reasons always exist behind every limitation; the "smart set" always accepts them.

WEIGHT AND BALANCE

Another very important part of the aircraft flight manual, although frequently published separately, is the weight-and-balance section. Ensuring that the aircraft weight-and-balance is within limits is an important part of flight planning, and is covered in that respective chapter. This section offers an understanding of the affect of excess weight and out of balance conditions on the performance and controllability of the airplane.

The problem with an overweight condition is not that the airplane cannot handle a little extra weight—usually it can if the truth be told. The problem is that the airplane might not be able to handle the additional stress. Normal category airplanes are certified to 3.8 Gs. An extra 100 pounds might not be a problem for the airplane to lift, but in a 3.8 G maneuver that extra 100 pounds becomes 380 pounds and the airplane might not be strong enough to handle the stress. The tail usually comes off first: not just the elevator,

THE BASICS

the whole tail. When an aircraft is overloaded the structural integrity is reduced. You might as well take a hacksaw and have a go at the spar, the engine mount, and the tail section—overweight stressing amounts to the same thing.

Out of balance can be more serious than overweight. A nose-heavy airplane is stable but inefficient. It will be very difficult to flair and might result in a hard landing, but seldom is it the cause of serious accidents. Tail heavy is very dangerous. The farther to the rear the center of gravity is, the less stable the airplane.

An unstable airplane will not recover from a stall without control inputs. If the center of gravity is too far back, the airplane will not recover from a stall—with or without control inputs; the nose will pitch up and stay there. The only way to recover from this kind of stall is to move the center of gravity forward, which is usually impossible in-flight. Test pilots do it by jettisoning ballast and popping drag chutes, but general aviation does not have that option. "Old time" pilots who found themselves in this situation used to climb out of the rear cockpit into the front cockpit—even if already occupied. This option might not be available either. It is much better to plan and check the balance properly prior to engine start.

REGULATIONS

You should thoroughly know the regulations that pertain to respective operations. For private pilots that means most of Federal Aviation Regulation (FAR) Parts 61 and 91. Fortunately, there aren't many regs; unfortunately they could be clearer and more specific. Because of this they are subject to endless interpretation and argument. For instance, in IFR conditions, a reserve of 45 minutes fuel at "normal cruising speed" is required. That's fine, but what is "normal cruising speed?" Most people assume that normal means whatever cruising speed is planned. I don't know that for sure; it might mean not long-range cruise, because that isn't "normal." Again, this might seem like nit-picking, but if an emergency declaration is caused by low fuel, this question might assume larger proportions.

The best defense against any brush with the FAA is a thorough knowledge of the regulations. Become inherently skeptical of the self-serving interpretations of other pilots; too often they will attempt to rationalize shortcuts by twisting regulations around to their point of view. Knowledge plus experience will result in the best interpretations. Discretion will minimize any problem.

CLEARANCES

Some pilots have trouble copying an instrument fight clearance, although with practice nearly everybody gets it sooner or later. Two things shorten the learning process: be prepared and know what is coming.

Prepared means having a piece of paper and a pen or pencil handy, which means always putting them in the same place. That is the only way they can be "handy." You

The System

don't need reams of paper and boxes of pencils, but you do need to develop a system that includes a place to write clearances and a place to always keep a pen or pencil. Then when you hear "34 Xray, clearance," you will be ready to write. Naturally, if you're busy with something else, like taxiing or an important checklist item, tell the controller to standby. It can wait. But when you *are* ready, be ready.

If you know what is coming in the clearance, and the order it will come, writing it down is very simple. The first item will be the routing, then the altitude, the transponder code, and the last item is usually the departure air traffic control (ATC) frequency. An additional remark, such as a void time is possible. I write them down line by line, according to the order given. Thus a typical clearance after I have copied it would look like this:

V91 PWL FPR
60 Exp 350 10 min
3457 119.85
1522

Four lines translate into a clearance: "34 Xray is cleared to SOP via V91 Pawling, flight planned route; maintain 6,000, expect Flight Level 350 10 minutes after departure; squawk 3457 departure frequency 119.85; clearance void if not off by 1522 Zulu." I would read back: "34 Xray cleared to SOP via V91 Pawling, flight planned route, 6,000, expecting 350 10 minutes after departure, 3457, 119.85, and void 1522." As long as the order is the same, the numbers are what count. Reading clearances ties up a frequency for a long time, so appropriate abbreviation is not only all right, it is good.

If ATC issues a routing you never heard of, write it down as you hear it, using whatever abbreviation comes in handy, and then make sure when you read it back that you say the words exactly as you heard them. This gives the controller a chance to issue corrections. You don't have to look up the complete new route before reading it back. Read it back, then look it up on the chart. If it still doesn't make sense, or you just can't find a fix, ask. Nine times out of 10 it will make perfect sense when you look it up, and in this way you do not tie up the frequency.

But whatever you do, *never* take off if you don't fully understand the clearance. It is much better to admit that you can't find a certain VOR or intersection or airway on the chart. (Verify chart currency and review the chart during flight planning.)

Some sort of shorthand system for copying a clearance is necessary, but I don't think you need to memorize any one system, nor does it have to be extensive. Create a personal system.

"THE SYSTEM"

These are the basic skills necessary to safely and routinely operate in the air traffic system. If none of this presents a problem, great. If you might be deficient in certain areas, practice, or take some instruction and become proficient. A mastery of these skills

THE BASICS

is expected of all pilots, regardless of the certificate; private, commercial, or airline transport.

The difference between a *pilot* and someone who simply *has* a pilot's license is a mastery of "The Basics."

2
Flight Planning

YOU CAN'T NOT FLIGHT PLAN. IT IS AS IMPOSSIBLE TO NOT FLIGHT PLAN AS IT IS to not think, and for the same reason. Even if the flight plan consists of nothing more than making sure the tanks are full, the sky is clear, and that you know how you are going to find the airport with the fuel range available, you have still flight planned. It isn't much of a flight plan, of course, but it is a plan of sorts.

The question, therefore, is not when to flight plan, but a flight plan method: What are the key elements of any good flight plan? How can these elements be adapted to fit both short and simple flights and long, complex flights? What is the purpose of a flight log? How is the plan actually used in the airplane? In short, how does a pilot plan a flight so it is completed safely, routinely, and as planned?

(If you want to learn more about flight planning for all operations—basic VFR, VFR on airways, IFR, high level, international—then check out a previous book of mine, *The Aviator's Guide to Flight Planning*, TAB 2438. This chapter covers the basics of good flight planning and the previous book explains how to actually do it for many operations.)

It might be easiest to start with what flight planning is not: flight planning is not merely filing an FAA flight plan nor is flight planning simply a process of completing a flight log (although each item is a part of good flight planning). Flight planning is everything you do before you actually take to the air: a complete description of how you intend to get from A to B, including a departure time, a route and distance, an altitude, a cruising speed, an estimate of en route time and fuel, provision for an alternate airport and reserve fuel, plus a weight and balance check, and a check of takeoff and landing runway lengths.

FLIGHT PLANNING

The flight log portion of the flight plan is a specific, detailed breakdown of the flight plan—a dress rehearsal, in effect, of the projected flight. Flight logs are not merely aids for student pilots, but are an important part of all but the most simple flight plans, and are essential in keeping tabs on the progress of the plan when airborne.

VFR AND IFR FLIGHT PLANNING: DIFFERENCES

In the most general sense, all flight planning is fundamentally alike. Every flight, regardless of the aircraft type or the flight rules under which the flight is to be conducted, requires that the same flight planning factors be taken into consideration: route, navigation, altitude, airspeed, winds aloft, fuel flow, alternate airports, and fuel reserves.

But the details of flight planning do vary considerably depending upon the type of operation, and in particular they vary depending upon the choice of VFR or IFR flight rules. Pilots operating under visual flight rules are probably most concerned with navigation (finding their way) while pilots operating under instrument flight rules are probably most concerned with fuel planning (making sure they don't run out of gas). (Navigation under IFR generally takes care of itself as a direct function of the airways clearance.)

These are important differences, but because this book is concerned mostly with flying as transportation, and because flying as transportation is something that is best done IFR, this chapter is most concerned with flight planning for instrument operations, along airways with IFR fuel reserves. In any case, the pilot who operates VFR can't go far wrong in observing IFR principles of flight planning.

This isn't to say that basic VFR flight using pilotage (flying by reference to geographical features) or dead reckoning (estimated headings and times) doesn't have its place as a purely recreational matter, but navigation using only pilotage and dead reckoning just isn't accurate enough or reliable enough for routine transportation.

Lindbergh crossed the Atlantic using only dead reckoning, but it didn't work very well, even for him (at one point he had to circle low over some fishermen to ask the way to Ireland), and he had no choice: long range electronic navigation systems did not exist. Pilotage and dead reckoning do have a place as a backup in case of total electrical failure. Flight planning in this chapter means flight planning along VOR airways, primarily for the purpose of operating on an instrument flight plan.

MAJOR COMPONENTS

Probably the easiest and most logical way to think about flight planning is to step back and look at the big picture. What is the intent of flight planning? *A determination, in advance, that the successful completion of the flight is a virtual certainty.* We can divide this task into two fundamental components: 1. How do we get there? 2. How much fuel is required? If we know, in some detail, exactly how we intend to go from A to B, and if we also know, in detail but allowing for contingencies, how much fuel it will take, then we can safely anticipate the successful completion of the flight.

Specifically, "How to get there" means a route and a safe altitude; "how much fuel" means time en route and fuel burned per hour, which means knowing the total distance, airspeed, groundspeed, and fuel flow in either gallons or pounds per hour.

For the simplest flight—good weather to a nearby airport over a well-known route—the flight planning might consist of nothing more than a mental review of route and altitude, estimated time en route, estimated fuel flow, and a check of the weather. Planning a more complex flight would be more detailed, with a written record made prior to takeoff, including the preparation of a flight log. But in every case flight planning boils down to route and fuel.

DESTINATION

I usually start flight planning with "there" in "how to get there:" the destination airport. One mistake pilots often make is to automatically select the airport nearest the ultimate destination. That is, if going to an office downtown, the nearest downtown airport is selected. Of if they are going to visit a factory in a small town they automatically go to the airport listed for that town. This is very often a mistake, for several reasons.

The problem with using the nearest airport to the downtown business district, or the only airport for a small town, is that often theses airports have less than optimum approach and landing aids. It doesn't make any sense to try and save time by using a close-in airport with only a VOR approach and fairly high minimums, miss the approach, and end up going to the big airport with the ILS and low minimums anyway.

Because the smaller airports usually don't have weather reporting nor forecasts, it is difficult to know what the weather is going to be at the time of arrival, and missing the approach is a very real possibility in many cases. Planning to use the major airport serving the area frequently saves time (no missed approach and diversion) and almost always reduces any risks associated with non-precision approaches to minimums.

Other reasons exist for not necessarily using smaller and more remote airports and secondary city airports: major obstructions on the approaches (tall trees, power lines, and smoke stacks); shorter runways; terrain-induced turbulence; and usually uncontrolled traffic. They might be nice places to visit on a weekend or attend an airshow, but in terms of routine transportation they impose an additional element of risk that is unnecessary and unjustified in most cases by the amount of time saved.

Even when the weather cooperates and a safe arrival is accomplished, using a smaller airport often does not save time anyway. Again and again in personal experience with corporate flying many passengers insist on going to the closest suitable airport to their destination when a larger and more complete airport was available a short distance away. Passengers might have to wait around for a taxi to show up, or wait for the party being met (with the transportation) to find the airport, or count on getting a rental car only to discover that none are available with the resultant loss of all time saved going to the close-in airport.

FLIGHT PLANNING

It's a fact of life that if the airport facilities are not all that great, that the ground facilities are usually not all that great either.

I am not saying that all small airports are bad: if the approach and takeoff minimums are satisfactory for the prevailing weather; if the approaches are reasonable and relatively straightforward, if the terrain is nonprecipitous, if the runways are long enough to provide a margin of safety, and if you have called ahead to check on the services, fine. Check everything out because it is an important area of flight planning.

My process to select a destination airport is to first identify all the airports near the ultimate destination and automatically eliminate those without an instrument approach. I locate the closest airport and check it out for suitability. If I have any question that I can't resolve, I eliminate that airport because there is almost always a large metropolitan or regional airport fairly nearby and I just don't want to take a chance of a questionable facility.

Advantages of using larger airports are so much greater than the disadvantages that they are generally the obvious choice. Selecting a major airport also makes the rest of the flight planning easier.

If you are nervous about using a large airport—unsure you can handle the traffic, or the controllers talk too fast—remember that pilots are not born with the ability to handle dense traffic. Learn how to handle high-density traffic by flying with an instructor into a terminal control area (TCA) or an airport radar service area (ARSA). (This is another area where an instrument rating, and the training that goes into it, can solve many problems.) Then, when you are ready to give it a try as pilot in command, pick a time and plan the trip carefully and give it a go. In all likelihood you will find flying to a major airport easier than to a smaller or less busy airport because normally everything is predictable and smooth at the major airports, and you have company to show the way. Pretty soon you'll wonder why you ever did it any other way.

PREFERRED ROUTES

After selecting a destination airport, the next step is to find the best routing. If you are going to one of the bigger airports around a major city, such as a large general aviation airport, the first place to check for routings is the "Preferred IFR Routes" section of the *Airport/Facility Directory*. This section shows routing between a variety of major city pairs. If flying from Boston to Chicago, for instance, the preferred routing from takeoff to the initial approach fix is listed. Preferred routes are still helpful when departing from a secondary airport. Look up the routing for the departure city that is nearest the actual place of departure and the arrival airport nearest the actual destination airport, and use as much of that routing as possible. You will have to figure out how to get on the route after takeoff and off the route to reach the destination, but the bulk of the en route portion will be the same.

Preferred routes offer some idea of what ATC expects you to file if one of the routes goes over or near your departure point. For instance, the preferred route between Boston

Preferred Routes

and Chicago goes directly over Albany, Utica, Syracuse, Rochester, and Buffalo. If departing from one of these cities, or from one of the many smaller airports around those cities, merely intercept the route that goes overhead, and the rest of the preferred route will still apply.

If you have done much instrument flying you have probably figured out by now that "preferred" really means "expected" or "normal." You can always file another route, but the preferred route is what you are going to get. In the air many changes are possible depending upon the traffic and the controller, but the initial clearance will almost always be for the preferred route. Naturally, if you have a good reason to go another way—"because it's shorter" is not a good reason, avoiding mountainous terrain or a line of thunderstorms would be—file it and explain in the remarks section of the flight plan. A short, simple phrase is sufficient: "Deviation around weather." But be prepared to wait until ATC can accommodate the request because special requests generally go to the end of the line.

Many preferred routes exists, but only a few are published. Practicality dictates that standard patterns of arrival and departure be established for all major airports, with appropriate transitions for the satellites. One way to learn the routes is visiting an air route traffic control center (ARTCC) and asking a supervisor for a briefing on how the arrivals and departures are handled in that center. You can learn so much about different aspects of the system from a visit to a center facility, but, in fact, it is not absolutely necessary to know the unpublished preferred routes. Knowing the routes ahead of time can assist flight planning, but it is not absolutely essential because you will be told.

Experience will reveal many of these routes after you have flown them several times. Flight service can also be a big help; they take flight plans all day long and have a very good idea of what the normal routings are.

In the absence of direct information, look at en route and area charts for the published holding fixes around the major airport in the area. Published holding patterns are usually part of the normal arrival patterns: where controllers stack arrivals when the approaches are saturated. (En route and missed approach holding fixes are not normally depicted on the en route and area charts because, in the case of en route holds, they are only needed rarely and in a random fashion, and missed approach holding fixes are shown on the approach plates.)

When all else fails, do not worry about getting the exact preferred or expected route. File the route that appears to make the most sense. Watch out for restricted areas and try to file using airways, with direct routes only as a last resort. You might be pleasantly surprised and get the route you file. This is great because it means that all preflight planning will be valid.

If you do get another route, you will have to make certain changes to the flight log en route, but that's just part of the game. It isn't as difficult as it sounds, but you want to avoid it if at all possible and that is why it is so important to try to anticipate the routing that will actually be assigned.

FLIGHT PLANNING

DISTANCE

The next step is to measure the distance between each fix and then add those distances to arrive at a total distance for the trip. Try to make this measurement of total distance as accurately as possible without being ridiculous about it. Take into account Standard Instrument Departures (SIDs) and known vectors and circuitous routings after takeoff, then measure the distances between direct fixes with a plotter. From the initial approach fix to the airport, measure the distance directly because there is no way to know which approach you will get or how you will be vectored.

Total distance is a main element of flight planning and much of what is done later will be predicated on distance. It is worth taking the time to get an accurate measurement.

ALTITUDE SELECTION

Many factors determine the amount of fuel required to complete a flight. One important factor is altitude because altitude determines the amount of power that can be developed, and power is directly related to airspeed and fuel flow. The forecast winds aloft also vary with altitude, generally increasing in velocity as altitude increases. The next logical step in the flight planning process is to determine a cruising altitude.

If weather never was a factor and the winds were always clam, the only factor in altitude selection would be distance; the longer the trip, the higher the altitude (up to the practical ceiling of the aircraft) because aircraft are more efficient at high altitudes. Let's look at an example based on the Beech Aircraft F33A Bonanza and a hypothetical 300-nautical-mile trip to examine these efficiencies.

Figure 2-1 is a portion of the F33A Pilot's Operating Handbook. Imagine a cruise power settings of 45 percent, and use the middle column for standard (ISA) conditions to note that at 3,000 feet, the F33A will use 9.6 gallons per hour for a true airspeed (TAS) of 132 knots. Going down the column, note that fuel flow is constant (because power is a constant 45 percent) while true airspeed increases approximately $1/2$ knot for each 1,000 feet of altitude.

How significant is this increase in airspeed with altitude? How much savings should be expected on a typical trip by cruising at, say, 9,000 feet instead of 3,000 feet? On a 300-nautical-mile trip, the 3,000 foot cruise at 132 knots will require two hours and 17 minutes; the 9,000 foot cruise at 137 knots will require two hours and 12 minutes, or five fewer minutes, for a fuel savings of 0.8 gallons. (To keep the example simple, it is assumed that the climb and descent legs cancel each other out; this simplification will not substantially affect the results.)

Five minutes is five minutes, and a gallon of fuel (we'll be generous) is a gallon of fuel, and if there is no reason not to go at 9,000 feet then you should. But it is not a large savings, and deviations around inclement weather and forecast winds aloft will probably outweigh the small gain in efficiency that comes with altitude.

From a practical point of view, which altitudes are available? The Beech manual lists cruise figures from sea level to 16,000 feet. In practice, the lowest IFR altitude available

Altitude Selection

CRUISE POWER SETTINGS

45% MAXIMUM CONTINUOUS POWER (OR FULL THROTTLE) 2100 RPM
3200 POUNDS

PRESS ALT.	ISA −36°F (−20°C)							STANDARD DAY (ISA)							ISA +36°F (+20°C)						
	IOAT		ENGINE SPEED	MAN. PRESS.	FUEL FLOW		TAS	CAS	IOAT		ENGINE SPEED	MAN. PRESS.	FUEL FLOW		TAS	CAS	IOAT		ENGINE SPEED	MAN. PRESS.	FUEL FLOW		TAS	CAS
FEET	°F	°C	RPM	IN HG	PPH	GPH	KTS	KTS	°F	°C	RPM	IN HG	PPH	GPH	KTS	KTS	°F	°C	RPM	IN HG	PPH	GPH	KTS	KTS
SL	26	−4	2100	20.4	57.6	9.6	127	132	62	17	2100	20.8	57.6	9.6	130	130	98	37	2100	21.2	57.6	9.6	132	127
1000	22	−5	2100	20.1	57.6	9.6	128	131	58	15	2100	20.5	57.6	9.6	131	129	94	35	2100	20.9	57.6	9.6	133	126
2000	19	−7	2100	19.8	57.6	9.6	129	130	55	13	2100	20.2	57.6	9.6	131	128	91	33	2100	20.6	57.6	9.6	133	125
3000	15	−9	2100	19.4	57.6	9.6	130	129	51	11	2100	19.9	57.6	9.6	132	127	87	31	2100	20.3	57.6	9.6	134	124
4000	12	−11	2100	19.1	57.6	9.6	131	128	48	9	2100	19.6	57.6	9.6	133	126	84	29	2100	20.0	57.6	9.6	135	123
5000	8	−13	2100	18.8	57.6	9.6	132	127	44	7	2100	19.3	57.6	9.6	134	124	80	27	2100	19.7	57.6	9.6	136	122
6000	5	−15	2100	18.5	57.6	9.6	133	126	41	5	2100	19.0	57.6	9.6	135	123	77	25	2100	19.4	57.6	9.6	136	120
7000	1	−17	2100	18.2	57.6	9.6	134	125	37	3	2100	18.7	57.6	9.6	135	122	73	23	2100	19.1	57.6	9.6	137	119
8000	−3	−19	2100	17.9	57.6	9.6	134	124	34	1	2100	18.4	57.6	9.6	136	121	70	21	2100	18.8	57.6	9.6	137	118
9000	−6	−21	2100	17.6	57.6	9.6	135	123	30	−1	2100	18.1	57.6	9.6	137	120	66	19	2100	18.5	57.6	9.6	138	116
10000	−10	−23	2100	17.3	57.6	9.6	136	122	26	−3	2100	17.8	57.6	9.6	137	118	63	17	2100	18.2	57.6	9.6	138	115
11000	−13	−25	2100	17.0	57.6	9.6	136	120	23	−5	2100	17.5	57.6	9.6	138	117	59	15	2100	17.9	57.6	9.6	138	113
12000	−17	−27	2100	16.7	57.6	9.6	137	119	19	−7	2100	17.1	57.6	9.6	138	115	55	13	2100	17.6	57.6	9.6	138	111
13000	−20	−29	2100	16.4	57.6	9.6	137	117	16	−9	2100	16.8	57.6	9.6	138	113								
14000	−24	−31	2100	16.0	57.6	9.6	138	116	12	−11	2100	16.5	56.6	9.6	136	110								
15000	−27	−33	2100	15.7	57.6	9.6	138	114																
16000	−31	−35	2100	15.4	55.6	9.3	135	110																

NOTES:
1. Full throttle manifold pressure settings are approximate.
2. Shaded area represents operation with full throttle.

Beech Aircraft Corporation.

Fig. 2-1. *This cruise power setting chart for the Beech F33A Bonanza shows that for a constant power percentage, 45 percent in this case, fuel flow will be constant for all altitudes while true airspeed increases with altitude.* For Educational Purposes Only. *Not to be used under any circumstances in the operation or maintenance of an actual airplane.*

(minimum en route altitude, or MEA) is 2,000 feet, and this is found only in flat country near the ocean. Generally MEAs will be higher.

The first step in determining an altitude "bracket" is to scan the route for MEAs; the highest MEA is the lowest altitude you want to file for. It's not a good idea to file for an MEA lower than that allowed for any one segment of the trip if you can possibly help it. If both you and the controller forget to climb to the higher MEA prior to reaching it, the outcome could be disastrous.

In theory, the top of the bracket will be either the service ceiling or the highest altitude shown in the performance charts (16,000 feet for the F33A). In practical terms, limit the top of the bracket to an altitude that allows a minimum 200 fpm rate-of-climb capability, ideally a 300 fpm climb capability. If the manual doesn't have a chart showing feet-per-minute remaining upon reaching a given altitude, it is easy to figure it out. Load the airplane to max gross takeoff weight and climb as high as possible at best rate-of-climb speed, on as close to a standard day as possible (a typical spring or fall day will be close to standard). When the rate-of-climb has dropped to a steady 200 fpm (or, better, 300 fpm), that is the *practical* service ceiling for your aircraft.

Limit the maximum altitude to 10,000 feet without supplemental oxygen, regardless of the climb capability of the aircraft. I personally feel the regulations are too loose to

Flight Planning

allow anything higher. (According to regulations, oxygen is required after 30 minutes between 12,500 and 14,000 feet, and all the time above 14,000 feet.) Even a healthy, young pilot who does not smoke will start to get short of breath above 10,000 feet, and most pilots will suffer at even lower altitudes. (An older pilot who smokes probably should not go higher than 8,000 feet without oxygen.) Reduce these limits by another 2,000 feet at night because eyes consume large quantities of oxygen and vision will definitely deteriorate above 8,000 feet.

To summarize, in a non-pressurized airplane, the maximum useful altitude range is 2,000 to 10,000 feet without supplemental oxygen and with oxygen, the highest altitude obtainable with a 200 to 300 fpm rate-of-climb capability remaining. For the F33A used in this example, that is approximately 16,000 feet for most conditions. Acceptance of this bracket simplifies the altitude selection process considerably.

Further, the range of altitudes is no lower than the highest MEA and normally (no oxygen or pressurization) no higher than 10,000 feet (8,000 at night). Also, winds aloft and weather are much more important factors in determining a cruising altitude than efficiency per se. Now is the time to decide which of these two remaining factors, winds aloft and weather, is the most important.

No one wants to flight plan into horrendous winds at altitude, nor intentionally give up a nice tailwind. Pilots like to go fast, and minimizing headwinds and maximizing tailwinds is part of the game. But headwinds and tailwinds are factors of time and money. Weather involves safety, and that has to take priority. I think even comfort takes some priority over speed. I don't like flying in or under clouds with turbulence, and most passengers hate it, regardless of what they say. Getting on top of any turbulence is very desirable and is worth a headwind penalty. Staying out of ice is worth a headwind penalty, and staying underneath towering cu's and heavy precipitation and turbulence and avoiding imbedded thunderstorms is worth sacrificing a tailwind.

The first step toward selecting an altitude should be a look at the weather, and especially any current pilot reports. Ask these questions first to select an altitude.

- Where are the cloud tops?
- Where is the ice?
- What kind of turbulence can be expected in the clouds?
- What sort of convective activity can be expected?
- How high to fly on top or above the ice?
- How low to fly and avoid towering cu's or stay beneath the ice?

In some cases, maybe even in many cases, none of these limiting factors will exist—the weather isn't always bad, just on the days when I have to fly. When none of these factors are limiting, it makes sense trying to maximize a tailwind by going as high as possible and trying to minimize a headwind by going as low as possible. But the rest of the time, forget about the winds aloft and select a cruising altitude that avoids the hazardous, adverse, and uncomfortable weather.

Power Selection

The more performance and altitude capability an aircraft has, the more complex the process of altitude selection is, which is one reason turbine-powered aircraft are frequently flight planned with computers. Fuel flows for turbine engines drop off dramatically with altitude, but the headwinds can also increase dramatically. When does the fuel saved get absorbed by going slower? This is a fairly tedious question to answer without a mainframe flight planning computer.

But the problem for nonturbine equipment is of an altogether different sort: picking the smartest altitude in terms of safety and comfort. This is actually more difficult because you can't put numbers in a computer to solve the problem. The way to solve the problem is to organize it, develop a list of priorities.

First, limit the choice of altitudes to a practical range. Second, examine operational considerations: weather, a smooth, ice-free, and, preferably, even cloud-free ride. If all altitudes are problem-free, then examine the winds: with a headwind go low, with a tailwind go high. It is great when it works out this way because it is like getting something for free: fly low one direction, ducking under the strong headwinds, then fly high the other direction to ride the strong tailwinds. But don't count on it and don't let it influence the choice of altitudes when other factors apply.

POWER SELECTION

For any given altitude, power can be set anywhere from the minimum necessary to sustain level flight to the maximum allowable as listed in the limitations section of the aircraft flight manual or as placarded or marked. However, for convenience in flight planning, manufacturers frequently provide power setting charts for several intervals of power from very low through very high. (The chart used in FIG. 2-1 for 45 percent is an example.)

The precise amount of power used is largely left to the pilot's discretion, and that choice is important enough to warrant a complete chapter: Cruise Control. The important point here is that the power setting chart (or schedule, as it is sometimes called) is where most of the performance information necessary for accurate flight planning is obtained, and until a decision has been made and a chart selected, no further flight planning can be done.

ESTIMATING TIME AND FUEL

The destination, route, and altitude have been selected and it is assumed that a cruise power schedule has been selected. The remaining important questions are time and fuel. Time depends on groundspeed, which in turn depends on airspeed and the winds aloft.

The airspeed part is easy because it comes right off the power setting table (FIG. 2-1). Groundspeed is more difficult.

The simplest and least accurate way to estimate groundspeed is to simply "eyeball" an average headwind or tailwind component, using the closest winds aloft altitude (3,000, 6,000, 9,000, etc.) as reported for a midpoint location. This is okay, but a little more effort will give much better results. Mentally averaging as above, but then getting out a

FLIGHT PLANNING

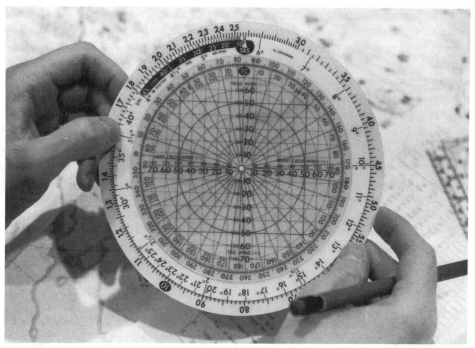

Fig. 2-2. *While electronic navigation calculators are becoming increasingly common in aviation, many professional pilots still prefer the circular "CR" mechanical navigation computer, shown here, for nav calculations because of their reliability, compactness, and familiarity.*

"whiz-wheel" (FIG. 2-2) to compute the actual component will result in a more accurate average groundspeed estimate than eyeballing. (For instance, with a TAS of 132 knots and a true course of 270, if the winds aloft are 315 at 50—45 degrees to the right of the nose—the headwind component will be 40 knots—not 25 or 30 knots as you might expect if you were to simply "eyeball" it.)

Maximize accuracy by interpolating the winds aloft forecasts: for the specific cruising altitude (not necessary if the cruising altitude happens to be one of the altitudes for which winds aloft are forecast—3,000, 6,000, 9,000, etc.) and for *each* station along the route of flight, and then compute an individual groundspeed estimate for each leg using the closest reporting point. This is more work than averaging and eyeballing, but it will result in better accuracy. With this kind of accuracy a pilot can make destination ETAs within a minute or two and total fuel required estimates within a gallon or two, even over max range distances.

The pilot who eyeballs a long trip and says it looks he should land with about an hour's worth of fuel could end up with anything from two hours worth of fuel to none. The pilot who determines a specific ground speed estimate for each leg and says he should land

2 hours and 57 minutes after takeoff with 16 gallons of fuel will probably land within a couple of minutes and a gallon or two of that estimate.

ALTERNATES

What about an alternate airport and the fuel to reach it? The regulations require an alternate for all IFR operations unless, for one hour before and one hour after the estimated time of arrival at the destination airport, the weather is *forecast to have a ceiling of at least 2,000 feet and visibility three miles or better*. In other words, the destination weather has to be better than normal VFR minimums (1,000 and three), or you must have an alternate airport. This includes any *possibility* that the weather will be less than 2,000 and three, which means the words or phrases often found in forecasts—chance of, occasional, or variable—also apply.

Personal Experience

Northeast United States forecasts often include something less than 2,000 feet broken or overcast or visibility less than three miles; I am in the habit of always filing an alternate airport. Besides, even good forecasts occasionally go completely belly-up, and airports are sometimes closed for reasons having nothing to do with the weather, the most common being disabled immovable aircraft. In many countries, an alternate airport is required to file an IFR flight plan regardless of the weather and that makes sense. Back-ups are always a good idea.

The main reason I always like to have an alternate airport is because I almost ran out of gas once when I did not have one and should have, even though it wasn't required. I was flying copilot on a Hawker-Siddeley 125-600. The weather was good in Boston with high overcast with good visibility and it was forecast to stay that way, and we had a short deadhead hop to Westfield, Mass, about 20 minutes away. We did not have full fuel, but we did have enough to fly the 20 minutes plus maybe 50 minutes reserve, so off we went. About half way there, just over Worcester, we ran into a sudden, very heavy, and completely unexpected snow squall. This was not much of a problem in itself because we had filed IFR anyway, but it did mean shooting an approach at Westfield.

The weather at Westfield was heavy snow but well above ILS minimums: still no problem because 50 minutes or so of reserve fuel was plenty for one ILS approach. Then the controller issued a hold over the beacon at Westfield: number three for the approach. *That* was a problem because we had enough fuel for one ILS approach, but waiting for two others was another matter. Then the really bad news—no aircraft was being cleared for any approaches because the hold was indefinite. We asked for an expected approach clearance time and he really didn't know, "Call it an hour, but it might have to be extended."

"That's no good, ask him for a clearance to Bradley," the captain said. The controller said Bradley (just south of Westfield) wasn't conducting any approaches either. We looked at each other and the captain said with some impatience: "Well, ask him for an approach back to Worcester."

This time the controller came back with, "Look, there aren't any approaches being approved anywhere in the area. I've got a guy lost in the clouds and we're trying to find him. Stand by."

What does he mean? Less than an hour away from logging multiengine glider time, there was nothing to do but get to work and look up the long-range and maximum endurance power settings and compute the time remaining.

Just as I picked up the mike to tell the controller that our time to fuel exhaustion was no more than 50 minutes, he said they had just found the lost pilot and had reopened the approaches. He cleared the number one holding aircraft for the approach, and we were, in turn, given vectors for the approach and ultimately landed safely, legally but shaken. "Never again," I said.

It took a series of unlikely events for this to become a problem, which is the way problems in aviation usually happen: tight on fuel to start with, the forecast was completely wrong, and someone was lost in the clouds. I learned that a series of unlikely events can happen and I have always planned on having enough fuel to get to an alternate, no matter what the weather is forecasted. If fueled for Bradley as an alternate, for instance, which is only a few minutes away from Westfield, that still would have been another 15 to 20 minutes of fuel, which would have helped a lot. If fueled for an alternate outside the immediate area, even better, no problem at all. I am not trying to blame the captain: I thought we had enough fuel, and "normally" it would have been okay; the captain listened to me.

Follow this rule of always having an alternate and forget about the "hour before and hour after" business—the 2,000 over and three, or is it 3,000 over and two?—forever. To be a legal alternate, the alternate must be forecast at or above alternate minimums at the time of arrival at the alternate airport. Alternate minimums are: 600 foot ceiling and two miles visibility if the airport has an ILS minimum of 200 and $1/2$; 800 and two for an airport with a nonprecision approach; and higher minimums for all other airports as published on the chart for that airport. There is no "hour before and hour after" business for an alternate, just "at the time of arrival at the alternate."

These minimums are deliberately higher than those required to complete an approach—600 and two versus the 200 and $1/2$—and this conservatism is intentional. It is meant to ensure that upon takeoff you will have somewhere to land. This extra margin means the alternate can be as much as 400 feet lower than the forecast ceiling and a mile-and-a-half less than the forecast visibility and still be at or above minimums, which means that unless the alternate weather drops way below forecast, an approach will still be possible.

The FAA does not specifically say so, but they don't want a pilot hopping from airport to airport, trying approaches until landing, or running out of fuel, whichever comes first. An unwritten rule: "When you miss an approach, don't fool around—go to the alternate and land." It doesn't make sense to carry extra fuel just to keep trying approaches at different airports. It never makes sense to try an approach at an airport other than the alternate with fuel intended for the alternate, unless the weather has changed substantially from that initially reported and forecast.

Take off and—barring very unusual circumstances like an emergency or a truly extraordinary change in the weather—plan on landing at one of two airports: filed destination or filed alternate because anything else probably means taking a chance.

Once in awhile, when a very powerful weather system dominates a large area, it can be difficult to find a legal alternate—weather forecast to be no worse than 600 and two. If an entire area is forecast to have low ceilings or poor visibility due to snow, heavy rain, or fog, sometimes the only airport without some mention of ceilings below 600 feet, or visibilities less than two miles, will be several hundred miles away. It hurts to carry fuel for an alternate so far away, especially when the weather has been low but still well above landing minimums all day and you are virtually certain you will make the approach at your destination.

And it *really* hurts when the nearest legal alternate is so far away that even with full fuel you can't go to the destination, proceed to the alternate and land with 45 minutes of fuel, which means a required fuel stop, even though the weather at the destination is above minimums. That doesn't happen often and that is the system; adherence has kept a lot of pilots from running out of fuel.

RESERVES

Obviously a pilot wants more fuel than merely enough to get to a destination and alternate. Another unpublished rule says always land with *some* fuel in the tanks. The regulations call for a reserve, at takeoff, of 45 minutes fuel at "normal cruising power." Most pilots interpret "normal cruising power" to mean the cruise power used for that particular flight. Therefore, if you flight plan using 55 percent cruise power, then plan on having 45 minutes of fuel remaining based on fuel flow at 55 percent power. If you flight plan for 75 percent power, then the required reserve will be proportionately higher.

Fuel flows for the Beech F33A vary from 15.2 gph at 75 percent power to 9.6 gph at 45 percent power. This results in an 11.4 gallon reserve at 75 percent, and 7.2 gallons at 45 percent power. These are minimums—there is no law against carrying more. The point is that 45 minutes of fuel is not an absolute.

I don't believe in tankering around a lot of extra fuel "for mama and the kids," but there is nothing wrong with figuring a generous amount of fuel as a personal "standard minimum reserve." I assume that if I really end up needing the 45-minute reserve that I am probably going to be shooting approaches and possibly climbing to holding altitudes after a missed approach, not merely cruising along at the higher altitudes.

I therefore assume a higher-than-normal fuel consumption during the entire 45 minutes. In a Falcon 20, for example, low altitude flying at high power settings results in about 3,000 pounds of fuel burned per hour, so I automatically use 2,500 pounds of fuel as a standard reserve. This is 50 minutes of fuel at 3,000 pounds per hour and more than 60 minutes at any of the "normal" cruise power settings: a nice even number well on the conservative side. Because it is already conservative, there is no need to "throw a little more on."

Unnecessary fuel is uneconomical, but, more importantly, it robs performance to

FLIGHT PLANNING

climb above ice, clear obstacles, or stop safely in an aborted takeoff or short field landing. Extra reserve fuel is fine, but there *is* a safety trade-off involved in carrying extra fuel; citing "safety" as the reason for loading excessive extra fuel is very often a cover-up for laziness and poor flight planning.

To continue the example of the F33A, start with 45 minutes of fuel at 75 percent power, which is 11.2 gallons for 45 minutes. Add two gallons for a missed approach and climb-out and then round up to 15 gallons for a standard reserve. Therefore, after figuring fuel requirements to reach a destination and alternate, automatically add 15 gallons to determine the fuel load.

The FAA also requires fuel for any "known" delays. Actual reported delays are very rare these days; most known delays are absorbed on the ground and do not require extra fuel. But I like to interpret the rule broadly and conservatively by thinking of "known" delays as meaning "likely" or "expected" delays. Thus I always assume there will be delays going into major metropolitan airports like LaGuardia, Kennedy O'Hare, Atlanta, Houston, Denver, and LA. The delays are usually fairly short, 20 to 30 minutes, but not unusual, so I put on an extra 30 minutes worth of fuel. If you fly to an airport that regularly has delays, fuel for that expectation.

Points to recall:

1. Fuel a trip to the destination airport, to an alternate airport, and land with a "fat" 45 minutes of fuel plus an allowance for known or expected delays.
2. Always have an alternate airport.
3. If you miss an approach at the destination, go directly to the alternate and land unless very unusual or extraordinary circumstances dictate otherwise.

When the inevitable happens and you do miss an approach at the destination airport, you are entitled to be slightly annoyed after landing at the alternate airport. But give yourself a little pat on the back for completing another safe flight, under difficult circumstances, routinely and uneventfully—that is what flying is all about.

AIRCRAFT FUELING

The basic flight planning is completed: destination, route, altitude, time, and fuel. An important number that comes out of this exercise is the total fuel required, which varies depending upon the conditions that prevail at the expected time of departure (in particular, depending upon the forecast winds aloft and the choice of an alternate). Therefore, the airplane cannot be properly fueled until the flight planning is complete.

This is contrary to common practice. Most pilots routinely fill the tanks after landing. It is usually more convenient to do it this way and having full tanks does minimize condensation, although condensation shouldn't be a problem in a frequently flown and thoroughly preflighted airplane. But the extra fuel carries a weight and performance penalty. Extra weight also carries an economic penalty because the extra fuel carried will require extra fuel to carry the weight, which costs money.

Flight Logs

The other way to fuel an airplane is to complete the flight planning, and then fuel up, bringing the total fuel load up to the required amount. A pilot can probably make bigger mistakes than overfueling an airplane, but it is still a mistake.

Fuel loading is also a critical element of weight-and-balance computations, therefore being in the habit of calculating the fuel prior to loading might prove valuable during the annual family vacation when every seat is filled.

FLIGHT LOGS

These general principles of flight planning for short trips over well-known routes with good weather are sufficient and provide reasonably accurate information for filing the FAA flight plan: how long the trip should be, a conservative alternate airport, and a substantial fuel reserve on board. Checking progress at the quarter, halfway, and three-quarters points for the effects of unforecast winds, changes in altitudes or routings, or the application of carburetor heat on our reserve fuel, is usually sufficient (again, for short trips in good weather over familiar routes) to keep us aware of any impending problems in time to do something about it.

On a longer trip, or a trip over an unfamiliar route or when the weather is questionable—and for the first 100 hours in any new airplane—take the planning a step further—prepare a flight log. There is nothing like the peace of mind that comes with having a good flight log to refer to and there is no better way to get to know an airplane than with a good flight log.

A flight log is a detailed breakdown of the flight plan, fix by fix, showing time and fuel estimates for each leg, which is updated with actual time and fuel burn en route. As you enter and compare actual time en route to estimated time en route and actual fuel remaining to estimated fuel remaining, the times will either be early, precise or late, and the fuel will either seem to "grow," stay the same, or "shrink." The figures and the trends are right in front of you—no surprises.

If the fuel consumption for the first leg is higher than expected, you are on alert. Maybe it was just a fluke and the next two legs are right on. But if the next leg turns out high also, then you know to do some checking.

Maybe the winds are stronger than forecast; maybe the airplane has something dragging (like a foot step that didn't retract or a baggage door that isn't completely closed); maybe the airplane is out of rig; maybe you forgot to lean the engine; maybe the power is set too high. Any of these factors would cause the fuel flow to be higher than estimated. Or maybe you just made a mistake reading the charts or a mistake in arithmetic. This is the time to find out, not 100 miles out with no good airports underneath you and the fuel remaining indicators bouncing around the bottom of the gauge.

A flight log does not have to be cluttered up with all the navigational information, like VOR frequencies or wind correction angles. The VOR frequencies can come right off the en route chart and the wind correction angles will take care of themselves. The main elements are distance, groundspeed, time, fuel burn, and fuel remaining. A sample flight log

FLIGHT PLANNING

```
FLIGHT LOG:    BOS    TO    BUF         ALT:  8000   POWER: 65%         FUEL LOAD
ROUTE: D--> MHT V490 UCA V2 BUF D-->    TAS:   163   FF:    13.3 GPH       64.0
```

			Trip Dist. 365		Time Off		Trip Time 2 Hours 53 Mins.		Trip Fuel 38.4	
Fix	Winds (-HW)	Est. GS	Dist.	Dist. Rem.	ETE	ETA	ATA	Actual GS	Fuel Req.	Fuel Rem.	Actual Fuel Rem.
BOS	/////	/////	/////	365	/////	///////| ///////	///////	/////	64.0	_____
MHT	-25	138	40	325	17	___	___	___	3.9	60.1	_____
CAM	-35	128	87	238	41	___	___	___	9.0	51.1	_____
UCA	-40	123	80	158	39	___	___	___	8.7	42.5	_____
SYR	-40	123	46	112	22	___	___	___	5.0	37.5	_____
ROC	-40	123	65	47	32	___	___	___	7.0	30.5	_____
BUF	-35	128	47	0	22	___	___	___	4.9	25.6	_____

Altn.		Avg. GS	Dist.	Rem.	ETE	ETA		Avg. FF	Req.	Rem.	
YYZ	------>	145	64	0	26			15.2	6.7	18.9	

```
IFR Fuel Planning   Gals.
    Destination:    38.4
    Alternate:       6.7
    Reserve:        15.0
    ---------------------
    Total:          60.1
    ---------------------
Fuel Onboard:       64.0
Difference:          3.9
    equals
    17 Mins @ 13.3 GPH
```

Fig. 2-3. *A flight log is a form of dress rehearsal for the flight to come, describing the proposed route, cruising altitude, time between fixes, fuel required between fixes, and fuel remaining over each fix. Blanks are provided for the entry of actual time and fuel values, an invaluable aid in keeping track of in-flight trends.*

for a typical trip from Boston to Buffalo, using performance figures for the F33A, is shown in FIG. 2-3.

This particular format works from left to right in a logical fashion, and the most important number—actual fuel remaining—is all the way on the right where it stands out and can be seen easily. The form does require sideways math. If you want to add or subtract directly under the appropriate number, modify the form to combine those columns. This will make the arithmetic easier, but might add clutter and confusion to the form.

I can't emphasize enough how important it is to maintain a good flight log. Every time I have had a problem with a flight it has been because I took off relying on approximations and rules of thumb instead of careful flight planning, and I didn't have a flight log to reveal early that a problem was developing. A minor problem can become a major problem when a pilot fails to realize it soon enough. Scrambling around trying to solve a problem at the last minute doesn't make sense anytime, especially not in an airplane—an airplane won't stand still. The time spent trying to solve the problem is more lost time and more time for the problem to develop. This creates pressure, and pressure is not conducive to good decision making. The best prevention is information and the best way to organize that information is with a flight log.

MAXIMUM RANGE TRIPS

Trips that reach the maximum available range pose a special problem: "Should you maximize range and fuel reserves, or plan to stop and not worry about it but losing time?" There is no right or wrong answer, but, I usually go with the planned stop. Most of the time when you try to complete an especially long leg you end up making a stop anyway when the winds don't cooperate, or ATC requests a less fuel efficient altitude, or the weather deteriorates at the destination and fuel reserves are not adequate. Then you end up getting the worst of both worlds: *flying slowly to begin with, then having* to descend, land, refuel, and replan and refile the remaining leg. If there is any question about the nonstop range it makes sense to plan a stop in the first place. (Chapter 3, Cruise Control, explains what to do when a stop isn't possible nor desirable.) A well-planned stop doesn't have to cost more than 15 or 20 minutes in ground time (versus 45 minutes to an hour for an unplanned stop), and the break is usually appreciated. The time spent on the ground for the turnaround can very often be recovered en route anyway—two short legs can be flown faster because fuel is no longer a factor.

This isn't an absolute. Planning long legs using long-range cruise techniques sometimes work and there is a certain satisfaction in maximizing the full range capability of an airplane. But it seems many times the planning does not work, so merely plan the stop.

WEIGHT AND BALANCE

The last part of flight planning is ensuring that the weight-and-balance is within limits. This might not seem like a logical part of flight planning, but remember that the purpose of flight planning is to provide a highly probable assurance that the flight can be successfully accomplished—weight-and-balance is a part of flight planning. Assuring that the weight is below maximum gross, that any zero fuel weight limitations are observed, and that the balance is within the normal envelope is an integral part of the flight planning scheme.

Federal Aviation Regulation (FAR) Parts 135 and 121 cover air taxi and airline operations, respectively, and require a weight-and-balance computation for every flight. The

Flight Planning

L-1011 LOAD MANIFEST

DATE	FLT. NO.	A/C NO.	DEP.	DEST.
2/27/90	AM7 024	N189HT	K BOS	LFPG

ITEM	WEIGHT	C.G.
ADJUSTED FA'S / BASIC WEIGHT ACM'S	2 4 2 6 2	7 4
FWD HOLD (C-1) BAGS ___ CARGO/WT. 2000	1 9 9	8 6
CTR. HOLD (C-2) BAGS 271 CARGO/WT.	9 5 0	2 7
AFT HOLD (C-3) BAGS 100 CARGO/WT.	2 5 0	1 1
BALLAST FUEL 0		
SUBTOTAL	2 5 6 6 2	9 8
ADULT PAX 320	5 4 4 9 0	4
CHILD PAX 0		
ZERO FUEL WT.	3 1 1 1 2 0	2
TAKE OFF FUEL 125.2 ✓	1 2 9 0 0 3	0
TAKE OFF WEIGHT 449.4 L	4 4 0 1 2 3	2
EST. FUEL BURN	1 0 8 0 0 0	
EST. LANDING WT. 354.0 5	3 3 2 1 2	3

C.G. ENVELOPE

WT.	FWD	ZFW AFT	AFT
240	12.8	30.3	34.0
260	12.7	30.7	34.0
280	12.7	31.0	34.0
300	12.6	29.3	34.0
320			34.0
325	12.6	28.5	
340	13.5	---	34.0
360	14.8	---	33.7
380	16.1	---	33.3
400	17.4	---	33.0
410	18.0	---	32.3
420	18.2	---	31.7
430	18.5	---	31.1
440	21.3	---	30.4
450	23.6	---	29.8

STAB TRIM

C.G.	SET
12	5.8
14	5.5
16	5.2
18	4.9
20	4.6
22	4.3
24	4.1
26	3.8
28	3.5
30	3.2
32	2.9
34	2.6

CARGO-BAGGAGE ADJUSTED WEIGHT TABLES

NUMBER OF BAGS		C-1 CARGO PIT*		C-2 CARGO PIT*		C-3 CARGO
INT'L	DOM.	FWD.	AFT	FWD.	AFT	**PIT
1-28	1-31	499.6	499.7	500.1	500.1	500.2
29-47	32-53	999.3	999.4	1,000.2	1,000.3	1,000.4
48-66	54-74	1,498.9	1,499.1	1,500.3	1,500.5	1,500.6
67-84	75-95	1,998.6	1,998.8	2,000.4	2,000.6	2,000.8
85-103	96-117	2,498.3	2,498.6	2,500.6	2,500.8	2,501.1
104-122	118-138	2,997.9	2,998.3	3,000.7	3,001.0	3,001.3
123-141	139-159	3,497.6	3,498.0	3,500.8	3,501.1	3,501.5
142-160	160-180	3,997.2	3,997.7	4,000.9	4,001.3	4,001.7
161-179	181-202	4,496.9	4,497.4	4,501.0	4,501.5	4,501.9
180-198	203-223	4,996.6	4,997.2	5,001.2	5,001.6	5,002.2
199-216	224-244	5,496.2	5,496.9	5,501.3	5,501.8	5,502.4
217-235	245-265	5,995.9	5,996.6	6,001.4	6,002.0	6,002.6
236-254	266-287	6,495.5	6,496.3	6,501.5	6,502.1	6,502.8
255-273	288-308	6,995.2	6,996.0	7,001.6	7,002.3	7,003.1
274-292	309-329	7,494.9	7,495.8	7,501.8	7,502.5	7,503.3
293-311	330-351	7,994.5	7,995.5	8,001.9	8,002.7	8,003.5
312-330	352-372	8,494.2	8,495.2	8,502.0	8,502.8	8,503.7
331-349	373-393	8,993.8	8,994.9	9,002.1	9,003.0	9,003.9
350-362	394-408	-----	-----	-----	-----	9,504.2
363-366	409-412	-----	-----	-----	-----	9,704.4

SHADED AREA: # BAGS AT 26.5 LBS. EACH (INT'L)
* STRUCTURAL LIMIT 18,000 LBS. MAX. (COMPARTMENT TOTAL)
** STRUCTURAL LIMIT 9,750 LBS. MAX.

FUEL

MINIMUM	1 2 5 2 0 0
TAXI/ADD'L	4 8 0 0
RAMP	1 3 0 0 0 0
LESS TAXI	1 0 0 0
T/O FUEL	1 2 9 0 0 0

L-1011 LOAD MANIFEST
COMPLETED BY:

NAME: D. Clausing

POSITION: F/O

ITEM #F20011   ATA OPS FORM IL (6/87)

Fig. 2-4. *Under Parts 121 and 135, weight-and-balance must be computed and verified within all limits prior to every takeoff. FOR EDUCATIONAL PURPOSES ONLY.*

major airlines use mainframe computers for their weight-and-balance checks; air taxi companies frequently develop weight-and-balance systems to simplify the task with charts that list all possible combinations of fuel, baggage, and passengers. Charter operators require either the first officer or flight engineer to manually check the weight-and-balance (FIG. 2-4). One way or another, a weight-and-balance check, showing loaded weight of the aircraft and the location of the C.G., is performed for every leg under Part 135 and 121.

As a result, an airline or charter flight rarely crashes due to improper weight and balance. But every year airplanes operating under FAR Part 91 do crash because of overweight or out-of-balance conditions, and these crashes are completely avoidable. It is fairly easy to see how they happen, though. Most of the time airplanes are automatically within weight-and-balance limits. Airplanes are designed to be in balance in normal use, and the typical load doesn't exceed the maximum weight limit. But it is too easy to have the exceptions sneak past.

For instance, if a pilot verifies that an airplane is within limits with full fuel and two passengers, and he has flown the airplane that way any number of times without a problem, he might not think to recheck the weight-and-balance when two passengers for a particular flight are big guys with heavy bags and both men sit in the backseat with the bags behind them. It just sneaks past until takeoff and the pilot notices the airplane does not handle correctly and is not performing well, or worse. Simply saying that "If there's any doubt, I always check," doesn't work because the pilot doesn't always "doubt" when he should. Include weight-and-balance as a regular part of flight planning and situations that exceed the limits will not sneak past.

FLIGHT PLANNING AS PLANNING

I flew several years for a corporation that went from a $100 million company to a $1 billion company in 10 years. During that time the flight department grew from a Cessna Citation and a Falcon 20 to two Citations, the Falcon 20, and a Falcon 50. When the chairman was asked in an interview for one of the leading financial journals what his secret was, he said simply, "There's no secret. Everything is done according to plan."

The company did not merely set a goal of $1 billion and tell everyone to work real hard, hoping it happened. A specific plan described exactly how to become a $1 billion company. The company put the plan into action, monitored its progress, and made changes as necessary to stay "on plan."

This is precisely what flight planning is all about. The significant advantage in the case of flight planning is having enough information to determine in advance whether the plan is achievable or not. Business plans are, unfortunately, not nearly as predictable, or everybody would be chairman of a $1 billion company.

Develop a flight plan, put it into action, monitor its progress, and make changes as necessary to ensure its successful and uneventful outcome. "Everything is done according to plan." It's the only way to fly.

3
Cruise Control

THE CRUISE PORTION OF FLIGHT USUALLY TAKES THE MOST TIME AND RECEIVES THE least attention. This is not all that surprising because when the aircraft has been leveled, trimmed, and leaned, there normally isn't too much to do other than keep track of the flight log, watch the VORs go by, check in and out with the controllers, and get an occasional update on the weather.

But that's not to say that the cruise portion isn't important, nor is it to say that cruise requires no thought or advance planning. The cruise portion of the flight might be less demanding than the takeoff, climb, descent, and approach segments, but cruise retains nuances and subtleties.

Cruise control, in the literal and technical sense, relates to the optimization of long-range cruise techniques; specifically, cruise control is the gradual reduction of power as fuel is burned off for the purpose of maintaining the most efficient angle-of-attack. This chapter covers certain aspects of long-range cruise techniques and describes why and how power and airspeed are reduced as weight diminishes, if maximum range is desired.

But the phrase *cruise control* is more widely and popularly used to mean merely the control of cruise power (and therefore airspeed and fuel flow) during the cruise portion of the flight. In this chapter, unless specifically referring to long-range cruise, cruise control is defined in the general sense of "control over cruise power."

EN ROUTE PERFORMANCE

Any excessive power setting that is higher than the amount of power needed to maintain altitude determines en route performance: speed, fuel burn, and "mileage." (Mileage

CRUISE CONTROL

is either "specific distance" or "specific range" in aviation. Instead of being expressed in terms of miles per gallon, as with cars, mileage is expressed in terms of nautical miles per pound of fuel. Multiplying by six (for avgas) will convert specific range to nautical miles per gallon; multiplying that result by 1.15 converts nautical miles per gallon to statute miles per gallon, which is what a car gets—miles per gallon. This is all part of the conspiracy to make flying seem more difficult than it really is.) The choice as to the percentage of power used is normally left to the pilot, and that choice is where cruise control begins.

EN ROUTE OPTIONS

The simplest and most common form of cruise control is to always use the same power setting. The pilot of an aircraft with a constant speed propeller might cruise at 2,400 rpm and 22 inches of manifold pressure, regardless of cruise altitude or winds aloft. For flight planning purposes the pilot would merely assume an average value for true airspeed and fuel flow—a rule-of-thumb. Like the broken clock that is correct twice a day, the rule of thumb will usually be correct for one or two altitudes, but most of the time the rule of thumb will give very inaccurate results.

This method has only one thing going for it—simplicity—and almost everything going against it. Accurate flight planning is nearly impossible with this simplification. The pilot can compensate for inaccuracy by carrying extra fuel, but that will not always solve the problem. In the absence of good flight planning, it is very possible that any problem that does arise won't surface until it's too late.

Because a rule of thumb is not only inaccurate but also inefficient, the pilot who uses this method also makes unnecessary fuel stops, often flies too fast, sometimes flies too slow, and frequently burns large amounts of fuel for very small gains in speed. It might be possible to "get away with" using rules of thumb in place of accurate cruise control, but you can do much better.

Cruise control pays dividends both in terms of safety and in terms of performance for the pilot willing to make the effort. The range of choices begins with long-range cruise (the slowest but most fuel-efficient power setting, resulting in the maximum nonstop range) and ends with maximum cruise (the exact opposite: maximum speed without regard for fuel consumed). Between long-range cruise and maximum cruise lies the "normal range," and several choices exist: low-power cruise, normal cruise, and high-power cruise. This chapter examines all three, plus a fairly new cruise concept known as "lowest cost cruise," neither fastest nor furthest, but cheapest.

LONG-RANGE CRUISE

Long range cruise (LRC) is seldom used in its pure form for the simple reason that long-range cruise is excruciatingly slow. It is so slow, in fact, that most general aviation owner's manuals don't even carry a schedule for it. In the absence of a long-range cruise schedule, approximate long-range cruise airspeed by substituting best rate-of-climb speed,

Long-Range Cruise

V_y: the speed that most efficiently converts power into altitude, V_y, is also very close to the speed that most efficiently converts power into distance.

The Beech F33A Bonanza previously used is an example of an aircraft without a specific long-range cruise power setting chart; V_y for the F33A is 96 knots indicated airspeed, therefore true airspeed at 6,000 feet would be 105 knots. Seventy-five percent power for the F33A results in a cruise TAS of 172 knots. In this case, flying at the approximate long-range cruise means giving up 67 knots of airspeed, and that is a steep price to pay for maximum range capability.

Certain situations demand that this price be paid. The most obvious situation is when confronted with a long, overwater route to an island airport with no fuel stops in between. With no possibility of making an unplanned fuel stop, and with no alternates available, the amount of fuel remaining at the destination airport becomes the key factor in determining the ability to compensate for navigational errors, delays, and weather, and LRC maximizes the amount of fuel remaining. In fact, on any long overwater route, regardless of whether alternates are available or not, LRC might be necessary. (It's hard to be too conservative over blue water.)

A less obvious but similar situation occurs when flying a route over land that lacks suitable fuel stops, either because of terrain or because of weather. For instance, the most direct route between Boston and Atlanta is approximately 850 nautical miles. A Bonanza at 55 percent power has an IFR, no-wind range at 12,000 feet of 860 miles—just barely enough. It might be worth a try anyway (assuming you have oxygen available for at least occasional use), as long as you have a good place to stop and refuel along the way if things don't work out as planned. But the problem with this particular route is that all the logical places to stop are in the mountains—not a great situation. So this leaves only two alternatives: fly out of the way, either to the east or west of the mountains to make a stop, or fly at LRC to conserve fuel and extend the range, and therefore extend the fuel reserves. (There is a third choice: go even higher than 12,000 feet. But that requires oxygen usage continuously for more than six hours.)

The advantage to using LRC as a strategy is that when you are sure the fuel stop will not be necessary, go ahead and push the power up a little and return to a normal airspeed. But when a pilot has flown off the direct route, that time is lost no matter what the winds or weather do.

Another time that LRC might make sense is when a given leg nonstop is possible at normal cruise speeds, but only with the help of a favorable winds aloft forecast. If the winds fail to materialize, it is a problem. A better idea might be to start out at LRC. If the winds are as good or better than forecast, go ahead and use normal cruise power. If the winds are worse than forecast, LRC will maximize the fuel reserve. You might still have to make a stop if the winds are extraordinarily adverse, but applying LRC techniques in this situation at least provides a "look-see" capability.

The key point here is that in each of these cases LRC provides an alternative to either not going at all, or having to make a fuel stop that is undesirable for one reason or another. Sometimes LRC is the solution to a problem, despite the tardiness.

CRUISE CONTROL

True LRC is based on flying at best lift over drag (L/D) speed. Best L/D is the speed where lift is greatest and drag is lowest, or, in plain English, the most efficient speed for the airplane. In practice, LRC schedules incorporate indicated airspeeds just slightly faster than best L/D. This reduces maximum range by only one percent (10 miles in a thousand), and ensures that the speed stays on the front side of the power curve. (Any decrease in airspeed below best L/D requires an *increase* in power to maintain level flight—the back side of the power curve—and that is *always* a negative.)

Unfortunately, best L/D speed, if not slow enough, gets even slower as the airplane gets lighter. This is because best L/D speed is a function of angle of attack, and the only way to maintain the optimum angle of attack in level flight is to slow the airplane as the weight decreases. Therefore, for absolute optimum LRC, the indicated airspeed for best L/D has to be reduced as fuel burns off. (This compensation for weight is what is meant by "cruise control" in the technical sense: control of indicated cruise airspeed for optimum range.) In other words, true LRC starts out as slow as an airplane can go, and then gets slower.

The good news, as a practical consideration, is that this form of cruise control generally is only necessary for turbine equipment where the weight difference between full fuel and minimum fuel can be substantial. The weight of a Falcon 20 business jet, for instance, can vary from 27,000 pounds at top-of-climb to 20,000 pounds at top-of-descent, a difference of 35 percent. Long-range cruise speeds for a loaded Falcon 20 at FL350 start out at Mach 70 (400 knots TAS), and gradually drop, as weight is burned off, to Mach 62 (355 knots).

For an airplane like a Bonanza, where the cruise weight might start at 3,400 pounds and drop to 3,000 pounds (only abut a 13 percent difference), cruise control in the technical sense is of negligible value, and the initial LRC airspeed can be held throughout the flight with very little range penalty. (Be grateful for small favors.)

Groundspeed is what really counts, not airspeed, and LRC groundspeeds can be reasonable with a decent tailwind. But with a headwind, LRC can be an exercise in frustration. If LRC TAS is 105 knots, and the headwind is 60 knots, the groundspeed will be 55 knots, and at that rate the range is going to be terrible no matter what the fuel flow is. In the extreme case, where the headwind equals the true airspeed, the groundspeed would be zero, and all fuel, no matter how low the consumption rate, would be wasted.

The way to compensate for the effect of headwinds is to increase the airspeed a given amount so that a balance is struck between the least amount of time spent penetrating the headwind and the most efficient airspeed. The exact amount to increase is best determined by flight planning computer, but Peter Garrison, who has an enormous amount of practical experience in this area, recommends (in his book *Long-Distance Flying*, Doubleday, 1981) that you increase the no-wind, LRC airspeed by one-fourth of the headwind component.

In the case of the Bonanza with a 60-knot headwind, that would mean increasing the indicated best L/D speed of 96 knots by 15 knots to 111 knots. (This rule doesn't apply to normal power settings because at those power settings the airplane is already going at least that much faster, and any additional increase will only decrease range.)

The key factor in achieving maximum range with reciprocating engines is proper leaning of the engine or engines. The difference in fuel flow between a properly leaned engine and one that hasn't been leaned at all can easily be 50 percent, enough to completely wipe out all the savings in going slow.

Many pilots are reluctant to lean the mixture, or at any rate are reluctant to lean it very much, because leaning might cause engine stoppage. The procedure for engine shutdown, is to bring the mixture back to the full lean position (actually the idle cut off position, but it looks like full lean) and pilots are afraid they will do that accidentally in-flight. Normal amounts of leaning don't get anywhere near the idle cut off position, but pilots are afraid of it.

Some pilots are also afraid of detonation, which they have heard can destroy an engine in a matter of seconds (which is true). Detonation *can* result from a very lean mixture and I don't want to minimize that danger. Detonation will most likely occur at high power settings—above 75 percent—and even then, only when the engine is running very hot.

Proper leaning under normal circumstances will not cause detonation and proper leaning will not cause an engine to quit. Leaning is not without dangers and should be done carefully and slowly to minimize those dangers. As long as the manufacturer's recommendations on leaning are followed (FIG. 3-1), there is no reason to fear the mixture knob. Without careful leaning, fuel flow will be horrible.

When you have a long way to go and no good way to break it up, LRC might be the answer.

VERY-LOW-POWER CRUISE

Many advantages of long-range cruise can be obtained at very low power settings (something around 45 percent power), but at much higher true airspeeds than true LRC requires (FIG. 3-2). Speeds at very low power settings will typically be 20 to 30 percent higher than LRC speeds (long range cruise might be called "extremely-low-power cruise"), and the range loss will seldom be more than 10 percent.

In any case, as a purely practical matter, most manufacturers do not provide LRC schedules for reciprocating-engine aircraft, which makes accurate long-range flight planning very difficult (when you need it the most), but manuals do provide very-low-power cruise information. This is a much more practical approach to maximum range than LRC; substitute 45 percent power (or something in that area) for true LRC, and it won't be far off.

LOW-POWER CRUISE

Low-power cruise is normally anything in the area of 50-60 percent power (FIG. 3-3). There are no hard-and-fast definitions for these labels. In this case, low-power cruise is simply the lower end of the normal cruise power region.

CRUISE CONTROL

Section IV BEECHCRAFT Bonanza F33A
Normal Procedures CE-674 and after

CRUISE

See Cruise Charts in PERFORMANCE Section

1. Cowl Flaps - CLOSED
2. Power - SET
3. Mixture - SET FUEL FLOW

LEANING USING THE EXHAUST GAS TEMPERATURE INDICATOR (EGT)

A thermocouple-type exhaust gas temperature (EGT) probe is mounted in the right side of the exhaust system. This probe is connected to an indicator on the right side of the instrument panel. The indicator is calibrated in degrees Fahrenheit. Use EGT system to lean the fuel/air mixture when cruising at 75% power or less in the following manner:

1. Lean the mixture and note the point on the indicator that the temperature peaks and starts to fall.
   a. CRUISE (LEAN) MIXTURE - Increase the mixture until the EGT shows a drop of 25°F below peak on the rich side of peak.
   b. BEST POWER MIXTURE - Increase the mixture until the EGT shows a drop of 100°F below peak on the rich side of peak.

CAUTION

Do not continue to lean mixture beyond that necessary to establish peak temperature.

2. Continuous operation is recommended at 25°F or more below peak EGT only on the rich side of peak.
3. Changes in altitude and power settings require the peak EGT to be rechecked and the mixture reset.

DESCENT

1. Altimeter - SET
2. Cowl Flaps - CLOSED

Beech Aircraft Corporation.

Fig. 3-1. *Manufacturer's recommendations on leaning, an example of which is shown here, should be followed carefully for maximum fuel economy and proper engine operation.* For Educational Purposes Only. *Not to be used under any circumstances in the operation or maintenance of an actual airplane.*

Low-Power Cruise

CRUISE POWER SETTINGS
45% MAXIMUM CONTINUOUS POWER (OR FULL THROTTLE) 2100 RPM
3200 POUNDS

PRESS ALT.	ISA −36°F (−20°C)							STANDARD DAY (ISA)							ISA +36°F (+20°C)						
	IOAT		ENGINE SPEED	MAN. PRESS.	FUEL FLOW		TAS	CAS	IOAT		ENGINE SPEED	MAN. PRESS.	FUEL FLOW		TAS	CAS	IOAT		ENGINE SPEED	MAN. PRESS.	FUEL FLOW		TAS	CAS
FEET	°F	°C	RPM	IN HG	PPH	GPH	KTS	KTS	°F	°C	RPM	IN HG	PPH	GPH	KTS	KTS	°F	°C	RPM	IN HG	PPH	GPH	KTS	KTS
SL	26	−4	2100	20.4	57.6	9.6	127	132	62	17	2100	20.8	57.6	9.6	130	130	98	37	2100	21.2	57.6	9.6	132	127
1000	22	−5	2100	20.1	57.6	9.6	128	131	58	15	2100	20.5	57.6	9.6	131	129	94	35	2100	20.9	57.6	9.6	133	126
2000	19	−7	2100	19.8	57.6	9.6	129	130	55	13	2100	20.2	57.6	9.6	131	128	91	33	2100	20.6	57.6	9.6	133	125
3000	15	−9	2100	19.4	57.6	9.6	130	129	51	11	2100	19.9	57.6	9.6	132	127	87	31	2100	20.3	57.6	9.6	134	124
4000	12	−11	2100	19.1	57.6	9.6	131	128	48	9	2100	19.6	57.6	9.6	133	126	84	29	2100	20.0	57.6	9.6	135	123
5000	8	−13	2100	18.8	57.6	9.6	132	127	44	7	2100	19.3	57.6	9.6	134	124	80	27	2100	19.7	57.6	9.6	136	122
6000	5	−15	2100	18.5	57.6	9.6	133	126	41	5	2100	19.0	57.6	9.6	135	123	77	25	2100	19.4	57.6	9.6	136	120
7000	1	−17	2100	18.2	57.6	9.6	134	125	37	3	2100	18.7	57.6	9.6	135	122	73	23	2100	19.1	57.6	9.6	137	119
8000	−3	−19	2100	17.9	57.6	9.6	134	124	34	1	2100	18.4	57.6	9.6	136	121	70	21	2100	18.8	57.6	9.6	137	118
9000	−6	−21	2100	17.6	57.6	9.6	135	123	30	−1	2100	18.1	57.6	9.6	137	120	66	19	2100	18.5	57.6	9.6	138	116
10000	−10	−23	2100	17.3	57.6	9.6	136	122	26	−3	2100	17.8	57.6	9.6	137	118	63	17	2100	18.2	57.6	9.6	138	115
11000	−13	−25	2100	17.0	57.6	9.6	136	120	23	−5	2100	17.5	57.6	9.6	138	117	59	15	2100	17.9	57.6	9.6	138	113
12000	−17	−27	2100	16.7	57.6	9.6	137	119	19	−7	2100	17.1	57.6	9.6	138	115	55	13	2100	17.6	57.6	9.6	138	111
13000	−20	−29	2100	16.4	57.6	9.6	137	117	16	−9	2100	16.8	57.6	9.6	138	113								
14000	−24	−31	2100	16.0	57.6	9.6	138	116	12	−11	2100	16.5	56.6	9.6	136	110								
15000	−27	−33	2100	15.7	57.6	9.6	138	114																
16000	−31	−35	2100	15.4	55.6	9.3	135	110																

NOTES:
1. Full throttle manifold pressure settings are approximate.
2. Shaded area represents operation with full throttle.

Beech Aircraft Corporation.

Fig. 3-2. *Very lower power settings, such as this 45 percent cruise power schedule for the Beech F33A Bonanza, provide most of the range of long-range cruise but with significantly higher true airspeeds.* For Educational Purposes Only. *Not to be used under any circumstances in the operation or maintenance of an actual airplane.*

(You might see the label "Economy Cruise" attached to power charts for 50-60 percent, which is as good a name as any, except aviation has been getting along for years with very little concern for economy, so I don't see any reason to start now. Reminds me of the boy who went up to the airplane owner and said, "Man, you must be rich to own an airplane." Airplane owner said, "No, but I used to be.")

Low-power cruise *is* economical, especially downwind. With a nice tailwind a pilot can very often regain airspeed lost at low power settings; in effect, trading the tailwind for fuel. Probably the best thing low-power cruise has going for it is the quietness and smoothness. Even noisy, shaky airplanes usually settle down around 55 percent or so, and that's an important consideration, both in terms of comfort and in terms of safety—noise and vibration create fatigue, and fatigue is one of the enemies of safety. Besides, common sense dictates little need for going fast to save 20 minutes, if it makes a pilot a nervous wreck, ears ringing and shaking all over. I guess a pilot could always use the 20 minutes saved to rest up, but somehow that doesn't seem like the right way to do it.

So, for reasons of comfort and safety alone a pilot might want to use low-power cruise as a standard cruise schedule. In any case, low-power cruise is a good normal cruise downwind because the wind does more of the work.

CRUISE CONTROL

CRUISE POWER SETTINGS
55% MAXIMUM CONTINUOUS POWER (OR FULL THROTTLE) 2100 RPM
3200 POUNDS

PRESS ALT.	ISA −36°F (−20°C)							STANDARD DAY (ISA)							ISA +36°F (+20°C)						
	IOAT		ENGINE SPEED	MAN. PRESS.	FUEL FLOW		TAS	CAS	IOAT		ENGINE SPEED	MAN. PRESS.	FUEL FLOW		TAS	CAS	IOAT		ENGINE SPEED	MAN. PRESS.	FUEL FLOW		TAS	CAS
FEET	°F	°C	RPM	IN HG	PPH	GPH	KTS	KTS	°F	°C	RPM	IN HG	PPH	GPH	KTS	KTS	°F	°C	RPM	IN HG	PPH	GPH	KTS	KTS
SL	26	−3	2100	23.0	68.8	11.5	140	145	62	17	2100	23.6	68.8	11.5	143	143	99	37	2100	24.2	68.8	11.5	145	140
1000	23	−5	2100	22.8	68.8	11.5	141	144	59	15	2100	23.3	68.8	11.5	144	142	95	35	2100	24.0	68.8	11.5	146	139
2000	19	−7	2100	22.5	68.8	11.5	142	143	55	13	2100	23.1	68.8	11.5	145	141	91	33	2100	23.7	68.8	11.5	147	138
3000	16	−9	2100	22.3	68.8	11.5	143	142	52	11	2100	22.9	68.8	11.5	146	140	88	31	2100	23.5	68.8	11.5	148	137
4000	12	−11	2100	22.1	68.8	11.5	144	141	48	9	2100	22.6	68.8	11.5	147	138	84	29	2100	23.2	68.8	11.5	149	135
5000	9	−13	2100	21.8	68.8	11.5	145	140	45	7	2100	22.4	68.8	11.5	148	137	81	27	2100	23.0	68.8	11.5	150	134
6000	5	−15	2100	21.6	68.8	11.5	146	139	41	5	2100	22.1	68.8	11.5	148	136	77	25	2100	22.7	68.8	11.5	150	133
7000	2	−17	2100	21.3	68.8	11.5	147	138	38	3	2100	21.9	68.8	11.5	149	135	74	23	2100	22.5	68.8	11.5	151	132
8000	−2	−19	2100	21.1	68.8	11.5	148	137	34	1	2100	21.6	68.8	11.5	150	133	70	21	2100	21.9	67.5	11.3	151	129
9000	−5	−21	2100	20.9	68.4	11.4	149	135	31	−1	2100	21.0	67.3	11.2	149	131	67	19	2100	21.0	65.6	10.9	149	126
10000	−9	−23	2100	20.1	68.0	11.3	149	133	27	−3	2100	20.2	65.8	11.0	148	128	63	17	2100	20.1	63.8	10.6	147	122
11000	−13	−25	2100	19.3	66.0	11.0	147	130	23	−5	2100	19.3	64.0	10.7	147	124	59	15	2100	19.3	62.0	10.3	145	119
12000	−16	−27	2100	18.5	64.0	10.7	146	126	20	−7	2100	18.5	62.1	10.4	145	121	56	13	2100	18.5	60.2	10.0	142	114
13000	−20	−29	2100	17.7	62.0	10.3	144	123	16	−9	2100	17.7	60.2	10.0	142	117	52	11	2100	17.7	58.4	9.7	139	110
14000	−24	−31	2100	16.9	59.8	10.0	141	119	12	−11	2100	16.8	57.9	9.7	139	112								
15000	−27	−33	2100	16.2	57.6	9.6	138	114																
16000	−31	−35	2100	15.6	55.6	9.3	135	110																

NOTES:
1. Full throttle manifold pressure settings are approximate.
2. Shaded area represents operation with full throttle.

Beech Aircraft Corporation.

Fig. 3-3. *Low power cruise schedules, such as this for 55 percent power in the Beech F33A Bonanza, typically fall in the 50-60 percent range, and are sometimes labeled "Economy Cruise." For Educational Purposes Only. Not to be used under any circumstances in the operation or maintenance of an actual airplane.*

NORMAL CRUISE

Most pilots call something in the vicinity of 65 percent power "normal cruise" and several manufacturers even label it as such (FIG. 3-4). Like most things labeled "normal," it can mean something and nothing at the same time. There is certainly nothing *abnormal* about using 65 percent power, but that doesn't necessarily make 65 percent power "normal" either. Nonetheless, as long as you understand that 65 percent power is not the only perfectly normal "normal" power setting, 65 percent power will suffice (more or less) as "normal cruise."

If one power schedule is going to be used all the time, normal cruise power is probably the one; it is fairly economical, fairly fast, fairly quiet, and fairly smooth; it is everything all other cruises are—both good and bad but in smaller amounts. It's a basic compromise.

For pilots who learned to fly at the Rule of Thumb School of Aeronautical Wizardry, and who appreciate simplicity above all else, normal cruise is the answer. Using one power schedule is nearly as easy as a rule of thumb and is much more accurate: economical downwind, fast enough upwind, and easy to use in any wind.

Low-Power Cruise

CRUISE POWER SETTINGS
65% MAXIMUM CONTINUOUS POWER (OR FULL THROTTLE) 2300 RPM
3200 POUNDS

PRESS ALT.	ISA −36°F (−20°C)								STANDARD DAY (ISA)								ISA +36°F (+20°C)							
	IOAT		ENGINE SPEED	MAN. PRESS.	FUEL FLOW		TAS	CAS	IOAT		ENGINE SPEED	MAN. PRESS.	FUEL FLOW		TAS	CAS	IOAT		ENGINE SPEED	MAN. PRESS.	FUEL FLOW		TAS	CAS
FEET	°F	°C	RPM	IN HG	PPH	GPH	KTS	KTS	°F	°C	RPM	IN HG	PPH	GPH	KTS	KTS	°F	°C	RPM	IN HG	PPH	GPH	KTS	KTS
SL	27	−3	2300	23.3	80.0	13.3	150	156	63	17	2300	23.9	80.0	13.3	154	153	99	37	2300	24.5	80.0	13.3	156	151
1000	23	−5	2300	23.1	80.0	13.3	152	155	59	15	2300	23.6	80.0	13.3	155	153	96	35	2300	24.2	80.0	13.3	158	150
2000	20	−7	2300	22.8	80.0	13.3	153	154	56	13	2300	23.4	80.0	13.3	156	152	92	33	2300	24.0	80.0	13.3	159	149
3000	16	−9	2300	22.5	80.0	13.3	154	153	52	11	2300	23.1	80.0	13.3	157	151	89	31	2300	23.7	80.0	13.3	160	148
4000	13	−11	2300	22.3	80.0	13.3	155	152	49	9	2300	22.9	80.0	13.3	159	150	85	29	2300	23.5	80.0	13.3	161	147
5000	9	−13	2300	22.0	80.0	13.3	157	151	45	7	2300	22.6	80.0	13.3	160	148	82	28	2300	23.2	80.0	13.3	163	146
6000	6	−15	2300	21.8	80.0	13.3	158	150	42	6	2300	22.4	80.0	13.3	161	147	78	26	2300	23.0	80.0	13.3	164	145
7000	2	−17	2300	21.5	80.0	13.3	159	149	38	4	2300	22.1	80.0	13.3	162	146	75	24	2300	22.6	79.0	13.2	164	143
8000	−1	−18	2300	21.3	80.0	13.3	160	148	35	2	2300	21.7	80.0	13.3	163	144	71	22	2300	21.7	76.3	12.7	163	139
9000	−5	−20	2300	20.9	78.1	13.0	160	145	31	0	2300	20.9	76.4	12.7	161	141	67	20	2300	20.9	73.9	12.3	161	136
10000	−8	−22	2300	20.0	76.2	12.7	159	143	28	−2	2300	20.0	73.8	12.3	160	138	64	18	2300	20.0	71.4	11.9	159	132
11000	−12	−24	2300	19.2	73.8	12.3	158	139	24	−4	2300	19.2	71.4	11.9	158	134	60	16	2300	19.2	69.1	11.5	158	129
12000	−16	−27	2300	18.4	71.3	11.9	157	136	20	−7	2300	18.4	69.0	11.5	157	131	56	13	2300	18.4	66.8	11.1	156	125
13000	−19	−29	2300	17.6	68.8	11.5	155	132	17	−9	2300	17.6	66.6	11.1	155	127	53	11	2300	17.6	64.5	10.8	153	121
14000	−23	−31	2300	16.9	66.4	11.1	153	129	13	−11	2300	16.9	64.4	10.7	152	123	49	9	2300	16.9	62.4	10.4	151	117
15000	−27	−33	2300	16.1	64.0	10.7	151	125	9	−13	2300	16.1	62.1	10.4	150	119	45	7	2300	16.1	60.2	10.0	147	113
16000	−30	−35	2300	15.5	61.9	10.3	148	121	6	−15	2300	15.5	60.0	10.0	147	115								

NOTES:
1. Full throttle manifold pressure settings are approximate.
2. Shaded area represents operation with full throttle.

Beech Aircraft Corporation.

Fig. 3-4. *Cruise power schedules in the 60-70 percent range, such as this for 65 percent power in the Beech F33A Bonanza, are typically labeled "Normal Cruise." For Educational Purposes Only. Not to be used under any circumstances in the operation or maintenance of an actual airplane.*

By using low-power cruise downwind and normal cruise upwind, a conservative *system* of cruise control is created because fuel economy is good in both directions and ground speeds will be reasonable and fairly constant. When the head wind blows against an aircraft, normal cruise compensates for some of the ground speed loss. When the tail wind comes from behind, back off and save a little fuel with only a slight loss in ground speed.

The same system may be used by pilots who like to go fast, only then normal cruise is used *downwind*. Pulling the power back to 65 percent might be hard for some, but the actual loss in ground speed will be slight because the wind is doing part of the work, and the fuel savings will be significant. Then, upon turning around and heading back into the wind, high speed cruise is an option. This system provides the satisfaction of high speeds without having to pay an excessive price in terms of fuel.

In summary then, normal cruise *is* a good normal cruise for simplicity and ease of use; it is also a good upwind cruise if optimization of fuel (within the normal range) is the goal, and, it is a good downwind cruise, if optimization of speed is the goal (again, within the normal range). "Swing cruise" would be a good name for it.

CRUISE CONTROL

HIGH-SPEED CRUISE

True high-speed cruise is approximately 75 percent power for most reciprocating engine aircraft (FIG. 3-5). Certain pilots use high-speed cruise all the time because they like to go fast, and that's as good a reason as any. After all, there might be a lot of reasons why pilots fly airplanes, but the only practical reason is because flying is generally faster than the alternative. If that's a pilot's reason to fly, high-speed cruise will get there the fastest.

But there are other reasons (and perhaps better ones) for going fast also. A Cessna 150 (or 152) is a slow airplane, and even Cessna Aircraft probably would not argue with that. It was, after all, designed as a trainer, and cross-country speed is essentially irrelevant in a trainer. But a Cessna 150 (and other similar airplanes) can be used for transportation, and as such perhaps should be operated at the highest cruise power possible. Fuel flows, in absolute terms, will remain quite low and any slower airspeed negates the advantages of air travel.

In fact, any airplane that is slow can benefit from using high-speed cruise. Pressurized airplanes are frequently mixed up with very high-speed company above 21,000 feet. High-speed cruise can be used to mesh better with the high-speed flow, which helps avoid vectors off the airway to let everybody else go by.

CRUISE POWER SETTINGS
75% MAXIMUM CONTINUOUS POWER (OR FULL THROTTLE) 2500 RPM
3200 POUNDS

PRESS ALT.	ISA −36°F (−20°C)								STANDARD DAY (ISA)								ISA +36°F (+20°C)							
	IOAT		ENGINE SPEED	MAN. PRESS.	FUEL FLOW		TAS	CAS	IOAT		ENGINE SPEED	MAN. PRESS.	FUEL FLOW		TAS	CAS	IOAT		ENGINE SPEED	MAN. PRESS.	FUEL FLOW		TAS	CAS
FEET	°F	°C	RPM	IN HG	PPH	GPH	KTS	KTS	°F	°C	RPM	IN HG	PPH	GPH	KTS	KTS	°F	°C	RPM	IN HG	PPH	GPH	KTS	KTS
SL	27	−3	2500	23.9	91.4	15.2	159	165	63	17	2500	24.6	91.4	15.2	163	163	100	38	2500	25.1	91.4	15.2	166	161
1000	24	−5	2500	23.6	91.4	15.2	161	164	60	16	2500	24.3	91.4	15.2	164	162	96	36	2500	24.8	91.4	15.2	168	160
2000	20	−7	2500	23.4	91.4	15.2	162	163	56	14	2500	24.1	91.4	15.2	166	161	93	34	2500	24.6	91.4	15.2	169	159
3000	17	−8	2500	23.1	91.4	15.2	164	163	53	12	2500	23.8	91.4	15.2	167	160	89	32	2500	24.3	91.4	15.2	171	158
4000	13	−10	2500	22.8	91.4	15.2	165	162	49	10	2500	23.5	91.4	15.2	169	159	86	30	2500	24.0	91.4	15.2	172	157
5000	10	−12	2500	22.5	91.4	15.2	167	161	46	8	2500	23.2	91.4	15.2	170	158	82	28	2500	23.7	91.4	15.2	173	156
6000	6	−14	2500	22.2	91.4	15.2	168	160	43	6	2500	23.0	91.4	15.2	172	157	79	26	2500	23.5	89.7	15.0	174	153
7000	3	−16	2500	22.0	91.4	15.2	169	159	39	4	2500	22.6	89.7	15.0	172	155	75	24	2500	22.6	86.7	14.5	172	150
8000	−1	−18	2500	21.7	89.4	14.9	169	156	35	2	2500	21.7	86.5	14.4	170	151	71	22	2500	21.7	83.6	13.9	171	147
9000	−4	−20	2500	20.8	86.5	14.4	168	153	32	0	2500	20.8	83.7	14.0	169	148	68	20	2500	20.8	81.0	13.5	170	143
10000	−8	−22	2500	20.0	83.7	14.0	167	150	28	−2	2500	20.0	81.0	13.5	168	145	64	18	2500	20.0	78.3	13.1	168	140
11000	−12	−24	2500	19.2	80.9	13.5	166	146	24	−4	2500	19.2	78.3	13.1	167	142	60	16	2500	19.2	75.7	12.6	167	137
12000	−15	−26	2500	18.3	78.2	13.0	165	143	21	−6	2500	18.3	75.7	12.6	165	138	57	14	2500	18.3	73.1	12.2	165	133
13000	−19	−28	2500	17.6	75.4	12.6	163	139	17	−8	2500	17.6	73.0	12.2	164	135	53	12	2500	17.6	70.6	11.8	163	129
14000	−23	−30	2500	16.8	72.9	12.2	162	136	13	−10	2500	16.8	70.6	11.8	162	131	49	10	2500	16.8	68.3	11.4	162	126
15000	−26	−32	2500	16.1	70.4	11.7	160	133	10	−12	2500	16.1	68.2	11.4	160	127	46	8	2500	16.1	66.0	11.0	159	122
16000	−30	−34	2500	15.4	68.1	11.4	158	129	6	−14	2500	15.4	65.9	11.0	158	124	42	6	2500	15.4	63.7	10.6	156	118

NOTES:
1. Full throttle manifold pressure settings are approximate.
2. Shaded area represents operation with full throttle.

Beech Aircraft Corporation.

Fig. 3-5. *Any cruise power schedule greater than 70 percent, such as that shown here for 75 percent power in the Beech F33A Bonanza, is normally thought of as being a high-speed cruise regime.* For Educational Purposes Only. *Not to be used under any circumstances in the operation or maintenance of an actual airplane.*

A Cessna Citation, for instance, is seldom operated at anything less than high-speed cruise. The Citation is a fast airplane compared to airplanes with propellers, but up in the higher flight level altitudes the Citation is still quite a bit slower than most other traffic. High-speed cruise minimizes the differences. It might make sense to use high-speed cruise in the face of accepting vectors off the airways to let faster aircraft pass.

High-speed cruise is also the cruise regime to use whenever the winds are "horrendous." What's "horrendous?" Whenever the headwind component (not the headwind itself, but the headwind *component*) reaches approximately 50 percent of normal cruise TAS, that's horrendous. It may even make sense when less than 50 percent, but with a normal cruise TAS of 140 knots and a headwind component of 70 knots, almost anyone would want to go as fast as possible to minimize that kind of impact on ground speed. When you have a comfortable definition of horrendous headwinds don't hesitate to use high-speed cruise when the winds meet or exceed that definition.

Sometimes the weather can force the use of high-speed cruise. I flew for several years out of an airport located right next to a river. From August to November the airport would fog over late at night almost every night and would stay that way until around 11 or 12 the next morning. Sometimes the difference between severe clear and zero-zero would be a matter of five minutes because that's how long it would take for the fog to roll up the hill from the river and cover the airport.

If a flight looked like a race with the fog, we'd push the power right up to high-speed cruise and try to save a night at the alternate. Note very carefully that it was not a hurried approach and landing, merely the cruise portion. If the weather closed in on the approach, that was just too bad. (Nothing in aviation is more dangerous than a rushed approach.)

Finally, the longer the trip, the more high-speed cruise makes sense; the shorter the trip, the less it makes sense. Only on longer trips does the difference in speed between high-speed cruise and normal cruise start to show up: on a trip of 1,200 nm, increasing the speed from 150 to 160 knots saves 30 minutes; on a trip of 100 miles, the same increase saves two and a half minutes. On a short trip, high-speed cruise is nothing but "sound and fury, signifying nothing." (*Macbeth*, Act V, Scene V.)

There are good reasons for at least occasionally using high-speed cruise rather than operate at conservative power settings. In general, the slower the airplane, and the longer the trip, the more sense it makes to use high-speed cruise.

MAXIMUM CRUISE

Maximum cruise is the upper limit for cruise power. For many airplanes, maximum cruise will be "redline;" others have a specific "maximum continuous power limitation;" still others have a "maximum normal operating power." (FIG. 3-6). Whatever the maximum normal power setting is, that is maximum cruise.

Maximum cruise could be called "emergency cruise," because that is actually what it is. Maximum cruise isn't normally used as a matter of course, therefore performance charts rarely indicate maximum cruise. Maximum cruise means maximum effort and maximum effort is usually reserved for those times when there is no option.

CRUISE CONTROL

> Section II
> Limitations
>
> BEECHCRAFT Bonanza F33A
> CE-674 and after
>
> ***AIRSPEED INDICATOR MARKINGS**
>
> | MARKING | KCAS VALUE OR RANGE | KIAS VALUE OR RANGE | SIGNIFICANCE |
> |---|---|---|---|
> | White Arc | 53-122 | 52-123 | Full Flap Operating Range |
> | White Triangle** | 152 | 154 | Maximum Speed for Approach Flaps |
> | Green Arc | 64-165 | 64-167 | Normal Operating Range |
> | Yellow Arc | 165-195 | 167-196 | Operate With Caution, Only in Smooth Air |
> | Red Line | 195 | 196 | Maximum Speed For All Operations |
>
> *The airspeed indicator is marked in IAS values.
> **Serials CE-884 and after, and CJ-156 and after.
>
> **POWER PLANT LIMITATIONS**
>
> ENGINE
>
> One Teledyne Continental Motors Corporations model IO-520-BA or IO-520-BB engine.
>
> *OPERATING LIMITATIONS*
>
> Take-off and Maximum
>   Continuous Power. . . . . Full Throttle, 2700 rpm
> Maximum Normal Operating Power
>   Serials CE-891 and after with 2-blade Propeller Installed and Serials CJ-156 and after with 2-blade Propeller Installed. . . . . . . Full Throttle, 2550 rpm
>
> Beech Aircraft Corporation.

Fig. 3-6. *The limitations section of the Aircraft Flight Manual, here a copy of the page describing airspeed and powerplant limitations for the Beech F33A Bonanza, as supplemented by placards and instrument markings, will define maximum cruise for any given airplane. For Educational Purposes Only. Not to be used under any circumstances in the operation or maintenance of an actual airplane.*

Maximum cruise is something to consider when facing a potentially serious problem that is getting worse. For instance, if all electrical power is lost except the battery, time is very critical, especially in instrument conditions. If the battery gives out while still in the clouds, navigation will be reduced to dead reckoning and all communications will be lost. Maximum cruise power will not affect the battery load, but it will minimize the time in the air. In this case, maximum cruise means maximum chance of finding visual conditions before the battery runs down.

Other situations could arise when maximum cruise might be appropriate: slow fuel leaks, loss of cabin heat, gradual deterioration of attitude or heading gyros (request vectors to VFR conditions if flying IFR), or almost any serious or potentially serious problem that can't be solved and seems to be getting worse.

Certain problems can worsen if power and airspeed are increased. Engine problems typically fall into this category (decreasing oil pressure, increasing temperatures, vibrations, propeller overspeeds) and generally worsen by increasing power. Maximum cruise is usually not the solution to structural problems either: a window cracking or a flap coming loose. The temptation with these problems is to get on the ground as fast as possible, but the solution is not maximum cruise. These problems usually worsen at high speed.

It is obviously impossible to list every situation where maximum cruise is or is not appropriate. That's why an airplane requires a pilot—to use his head and make a decision based on the particular situation that arises. The important point here is that maximum cruise might be part of the solution for any problem that is time-critical.

LOWEST COST CRUISE

Lowest cost cruise (LCC) is an interesting concept. The idea originated with the airlines. When the fuel crunch came in 1973 and jet fuel went from 15¢ and 16¢ per gallon to $1 per gallon, the airlines responded logically by flying slower to conserve fuel. The problem was, as they flew slower, fuel costs did go down, but maintenance costs, crew expenses, and engine reserves *went up*. The reason was that slower meant more time on the airplane, and that cost something too.

Airlines discovered they could only go a certain amount slower and still achieve a net savings—any slower and the hourly costs exceeded the fuel savings. The compromise speed became lowest cost cruise.

Crew expenses are a very large factor in the LCC equation because crews that work more must be paid more. Privately operated aircraft don't have crew expenses, of course, but two other major factors that determine LCC—maintenance and a reserve for overhaul—do apply. Determining what LCC is for an airplane is not hard. All it takes is a performance manual, a calculator (a personal computer with spreadsheet software is ideal), and a little bit of time.

The first step is to determine what the hourly fuel costs for a typical cruise altitude, over a range of power settings. The relationship between power and cost will be essentially constant for all altitudes, so only one altitude is necessary. Determining fuel costs is

Table 3-1. Direct Hourly Cost, 45 Percent to 75 Percent Power (Typical).

Power Percentage	45	55	65	75
1. Fuel @ $2.00/gal	$19.40	$22.90	$26.60	$30.50
2. Maintenance	$12.00	$12.00	$12.00	$12.00
3. Overhaul	$6.00	$6.00	$6.00	$6.00
4. Total	$37.40	$40.90	$44.60	$48.50

a simple matter of looking up hourly fuel consumption figures in the performance manual and multiplying those figures by an average cost for fuel. (If using a spreadsheet, put the fuel cost in a separate cell to easily recompute as fuel prices vary.)

TABLE 3-1, Line 1, shows how this would look for a typical single-engine aircraft, as power varies from 45 to 75 percent.

The next step is to determine the hourly cost for scheduled maintenance. If an airplane gets regular 100-hour inspections, this is also a fairly easy number to calculate. Make a best cost estimate of a typical 100-hour inspection, divide that figure by 100 and the result is maintenance cost on an hourly basis. (If an annual inspection is performed, scheduled maintenance is not a factor. Presumably the "annual" will cost the same no matter how fast or slow an airplane is flown.) TABLE 3-1 shows this figure for typical single-engine airplane on Line 2. Note that it is a constant figure for each percent of power; a 100-hour inspection should cost essentially the same regardless of the amount of power used during those preceding 100 hours, within the normal range of power settings.

Unscheduled maintenance is not included for the simple reason that unscheduled maintenance is only indirectly related to hours in operation because flying faster or slower doesn't affect the "breakage" rate directly or predictably. For purposes of determining lowest cost cruise, the relationship between power and unscheduled maintenance is too vague to be useful.

The final hourly cost figure is a reserve for overhaul. Despite the fact that an allowance for overhaul is not an hourly, "out-of-pocket" cash expense, this chicken will come home to roost sooner or later either in the form of a bill for engine overhaul, or, if not in an actual cost, then in the form of increased depreciation at the time of sale. So it has to be accounted for to obtain a true picture of hourly costs. A local mechanic or FBO should be able to provide a figure regarding overhaul cost and should also know what the recommended time between overhauls (TBO) is. Coming up with an hourly figure is a simple matter of dividing estimated overhaul cost by the TBO. If an engine won't make it to TBO, use a best guess of how long it will last, but don't count on it going over.

TABLE 3-1 lists the hourly reserve for overhaul on Line 3. It is also a constant figure that does not vary with power. I know it is possible to argue that an engine operated at 55 percent power might be more likely to reach TBO than one operated at 75 percent power, but the relationship is debatable—use a constant figure.

Lowest Cost Cruise

Table 3-2. Lowest Cost Cruise (LCC).

Power Percentage	45	55	65	75
1. True Airspeed, 6,000 FEET	135	148	161	172
2. Cost Per Hour	$37.40	$40.90	$44.60	$48.50
3. Cents/nm	27.7¢	27.6¢	27.7¢	28.2¢
4. Per 100,000 nm	$27,703.70	$27,635.14	$27,701.86	$28,197.67
5. Lowest Cost Cruise		= = = = = =		
6. Amount Above LCC	$68.57	$0.00	$66.73	$562.54
7. Percent Above LCC	0.2%	0.0%	0.2%	2.0%

The sum of these figures, shown in TABLE 3-1 on Line 4, is total hourly cost for fuel, maintenance, and overhaul. As power increases, the cost increases, because fuel costs go up. But that's only half the story.

TABLE 3-2 is the other half of the story—notice how much actual work is produced by the fuel expended. Line 1 is just a simple listing of true airspeed (TAS) for each power setting, taken straight from the performance chart of a typical airplane. Line 2 repeats the total hourly costs from TABLE 3-1. Line 3 shows what each mile actually costs, to the nearest 10th of a cent. This was obtained by dividing Line 2 (cost per hour) by Line 1 (true airspeed). The result is cost per nautical mile.

Read across line 3 and realize that this particular airplane costs $.277 per mile at 45 percent power, $.276 per mile at 55 percent power, $.277 per mile at 65 percent power, and $.282 per mile at 75 percent power. Fifty-five percent power seems to be the cheapest, but the differences, on a per mile basis, are so small as to be virtually meaningless.

The differences might be more meaningful if the cost per mile was carried out to seven places and multiplied by 100,000 miles. Line 4 shows what the difference would be for 100,000 miles, which might sound like a lot but at 172 knots that is only 581 hours. Even here the difference is very slight, but it is easy to see that 55 percent is the cheapest cruise (LCC), that it costs only $66.73 more per 100,000 hours to use 65 percent power (Line 6), and that it costs $68.57 to use 45 percent power, presumably for the privilege of going slower. Stepping up to 75 percent power costs $562.54 more than 55 percent—2 percent increase over LCC (Line 7).

This means, for all practical purposes, that the cost is the same to fly this airplane at 65 percent power as it is at 55 percent power, it costs only 2 percent more to fly at 75 percent than it does at 55 percent, and there is no cost advantage at all in flying *slower* than 55 percent. (Operational reasons for flying slower than 55 percent might exist, but it will cost money to do so).

The only way to determine LCC for an airplane is to go through the analysis above. If an airplane is easy to maintain and relatively cheap to overhaul, LCC will be biased toward the low-power side.

On the other hand, a very sophisticated airplane that costs a lot to maintain and has an engine with a low TBO and high overhaul cost will be biased toward the high-power end—

CRUISE CONTROL

might as well fly it fast because it is going to cost more money otherwise. There is no way to know what an actual LCC is without plugging in reliable numbers. The results might be surprising.

It probably goes without saying, but when renting aircraft by the hour, LCC will almost certainly be high-speed cruise, or as fast as possible. When paying by the hour, faster means less cost. Even with a "dry" lease (pilot buys the fuel), the hourly charge will usually be high enough that the time and money saved going fast will more than compensate for the extra fuel purchased.

Do not abuse the rental airplanes to save a little money, but from an economic point of view there is no reason to use anything other than the highest normal power setting. Anything else is a gift to the owner.

None of this is meant to suggest that fuel conservation is not an important consideration on both a moral and a practical level. Oil is a finite resource, and when it is gone, it is *gone*. But if economics is the science of the allocation of scarce resources, lowest cost cruise implies that engines and airframes are scarce resources and also worthy of conservation.

Lowest cost cruise represents the optimum economic point of balance, which means, in simple terms, that operating an aircraft at LCC, not only saves a little money, but also strikes the best balance in terms of total conservation of resources. Even if the financial savings are not great, it's still a move in the right direction.

CONCLUSION

Accurate cruise control is not as difficult or complicated as this fairly detailed discussion might indicate. Neither is it so inconsequential as to not be worth a little thought and analysis. Even the simplest program of cruise control using one cruise power schedule for all altitudes and winds will pay dividends in terms of more accurate flight planning and more economical operation than any rule of thumb. Even greater returns are available in the form of improved efficiencies, extended ranges, and maximized economies, with just a little more effort.

In almost every professional cockpit, the most beat-up book is the cruise performance manual.

In a two-pilot airplane this manual is often kept in a pocket behind the right seat where the captain can reach it easily. The captain reaches for it after the airplane has leveled off to set initial cruise power. If there is an altitude change, the arm goes out again to get a new power setting. Even if the altitudes don't change, usually the arm goes out one or more times to get updates as either the outside temperature changes or fuel is burned off.

In a three-pilot airplane, the flight engineer normally keeps "the book" and provides the pilot with the necessary cruise information (fuel flow, Mach number, indicated airspeed, exhaust pressure ratio) often written on little sticky notes to post on the panel.

Nobody thinks twice about it because cruise control is a normal part of en route transport flying like talking to controllers and tracking VORs.

Conclusion

Good cruise control is the best sign of a well-defined cockpit and indicates a deliberate and thoughtful approach to flying that is one mark of a good pilot.

It isn't that good cruise control is the most important thing in operating an airplane, but it does mean that if this is being done right, then probably everything else is being done right too.

4

Approaches

EVERY TIME YOU TAKE AN AIRPLANE UP, YOU ARE COMMITTED TO BRING THAT airplane down. Remember the feeling on your first solo? "I know I can do it, which is good. I have to because there's nobody else in the airplane." If you are regularly using an aircraft for transportation, then you are probably also regularly flying under instrument conditions, which means that in many cases prior to that landing there is a commitment to an instrument approach in order to bring that airplane down.

Most passengers don't like approaches because they are at the mercy of the pilot and there is nothing they can do about it. A senior VP was riding in the back of a chartered Cessna Citation because the company's Citation was down for maintenance. It was the middle of winter and the weather was terrible: lots of ice, bumpy, low ceilings, the works. The pilot was shooting an approach to Lebanon, New Hampshire, the corporate flight department base. The "long" runway at Lebanon—the one that is never aligned with the wind, but the one you have to use anyway—is only 5,500 feet long, and on this particular night it was also covered with ice. Mr. VP hollered up to the pilot:

"Captain, does this Citation have reversers?"
"No, Sir."
A short pause.
"Captain, does this Citation have a drag chute?"
"No, Sir."
A long pause.
"Captain, does this Citation have any more Scotch?"

Passengers are at the pilot's mercy and it is a tremendous responsibility to do the best possible job on every approach. This chapter will discuss many approaches, but not with

APPROACHES

the intent of trying to teach how to fly them—that can only be done in an airplane or simulator. Instead, the chapter will cover little things that can make the difference between a shaky approach and a smooth approach and the difference between an approach that warrants passenger trust, and an approach that doesn't. The chapter begins with a discussion of the most demanding of approaches, the circling approach (when the inbound approach course is not lined up with the landing runway), then moves on to nonprecision approaches (both VOR and NDB), and precision approaches (the only approaches truly worthy of the name).

CIRCLING APPROACHES

A circling approach is any approach that the final approach course and the runway differ by more than 30 degrees, which is an approach that is not more or less straight in and therefore requires "circling" in order to be properly aligned. Circling approaches are among the most critical maneuvers in aviation. Many corporate flight departments require pilots to use Category D circling minimums, regardless of the actual aircraft category. Some flight departments specifically forbid night circling approaches; some forbid all circling approaches. (The airline I fly for prohibits circling approaches unless the weather is VFR, which really makes them visual approaches.) Circling approaches can be safe, but only if done properly and with scrupulous regard for their limitations. In other words, they're as safe as the pilot flying them.

Optimum

One of the best circling approaches I've ever seen was flown by a guy named Dave Naylor, ex-Air Force F-102 Interceptor pilot. I've seen other pilots shoot a lot of good approaches—approaches to ILS minimums, approaches with turbulence and ice, approaches in rain so heavy you had to yell to be heard, and approaches in the simulator with most of the systems out and the rest on fire—but this approach was still one of the best.

Well, sure, a former fighter pilot *ought* to be able to shoot a good circling approach. You're right, except all I knew about Dave Naylor at this point was that he was one of the least likely "fighter pilot" types I'd ever known. Dave was the kind of guy who thought it would be fun to spend two or three hours getting the coffee pot nice and shiny. His idea of a wild and crazy time was to stir his beer until all the bubbles were gone. But that didn't mean Dave couldn't fly an airplane. (I guess it merely meant he liked clean coffee pots and flat beer.)

What made this particular approach (a circling VOR DME approach to Hilton Head, South Carolina, FIG. 4-1) such a good approach was not that the approach was very tough. It *was* a tough approach, circling over the airport right on the deck at circling minimums, but it wasn't *that* tough. What made his approach so good was that he made it look so easy.

Dave was flying the Hawker-Siddely 125 and I was running the checklist, watching the clock, and occasionally looking outside. I didn't see anything outside all the way down the final approach segment until reaching the minimum descent altitude (MDA). At the

Circling Approaches

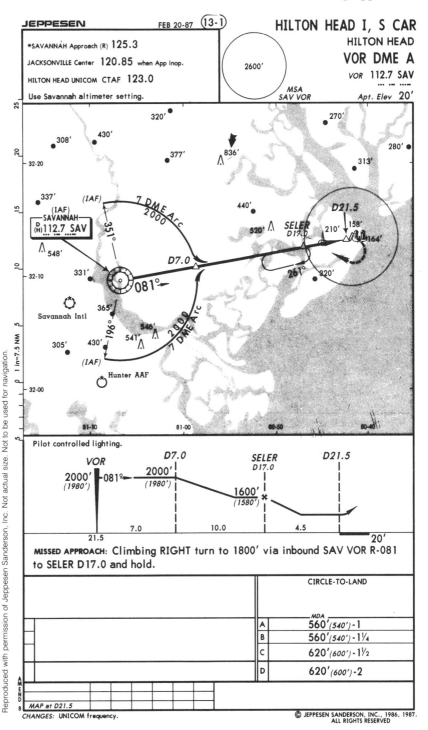

Fig. 4-1. *This VOR/DME A approach to Hilton Head, South Carolina, is typical of circling approaches. Note that no runway identifier is included in the approach heading and no minimums are listed in the bottom left-hand section, the section for straight-in minimums.*

APPROACHES

MDA the aircraft broke out under a solid overcast with good visibility underneath, but without a foot to spare: "up" meant back into the clouds and "down" meant descending below the MDA. I looked straight down and saw the airport right underneath. I called it out, just on principle, but assumed Dave would call for the missed approach anyway because the airplane was passing the airport. I expected another shot at the approach, to try to get down to the MDA a little quicker to set up for a right downwind (FIG. 4-2).

But Dave continued more or less straight ahead exactly at circling minimums—no missed approach, no climb-out, no call outs, just straight-and-level—with the clouds skimming the top of the airplane and the altimeter glued to the MDA.

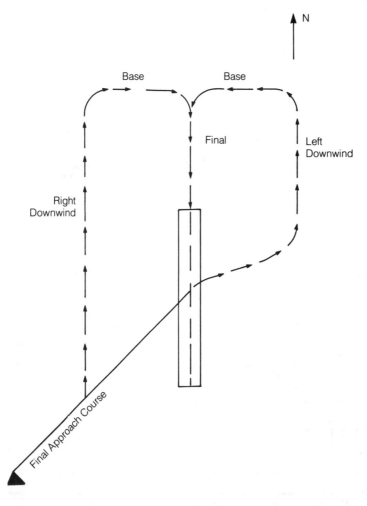

Fig. 4-2. *Two possible circling patterns for a runway that lies at an angle to the final approach course.*

Circling Approaches

Finally, after what seemed like approximately 30 minutes, but was actually only 15 or 20 seconds, Dave started a 25-degree-bank to the left, turning downwind, and I realized for the first time that he was going to circle back to the runway, wherever that was. Dave continued on that heading for about 30 more seconds, banked 30 degrees to the left and started back around for the airport.

Through 90 degrees—where the base leg would be—the runway came into sight and he called for the gear and flaps and the rest of the landing checklist, then lined up on final, slowed to exactly "bug" speed (the reference speed on final), touched down, put the spoilers out, went to lift dump (90 degrees of flaps), applied maximum braking, and taxied off the runway without so much as a word.

The landing distance at Hilton Head is only 4,000 feet long, counting the overrun, and there are trees at each end. That's not very much runway for a sweep-wing jet with a final approach speed of 130 knots.

Execution of the approach and landing was flawless. There was no confusion when the runway was first sighted, no diving for the runway, no skimming the treetops or groping around in the murk—merely a beautiful approach that was so smooth and controlled the passengers probably never thought twice about it.

That is one of the ironies of aviation: do something really well and nobody notices. The smooth, uneventful flight that proceeds exactly as planned, even under difficult circumstances, is boring and unimpressive to the passengers. But that is what every pilot is looking for. Everybody judges pilots on their landings, although the smoothness of a landing doesn't mean a thing in and of itself. But what *is* hard to take is when passengers don't even notice when everything is done just right. But that is the paradox: a "no comment flight" is truly the highest compliment.

Descent to the MDA

The Hilton Head approach story illustrates many of the important points about non-precision and circling approaches. First, get down to the MDA as soon as possible after crossing the final approach fix (short of dumping or diving the airplane). (If Dave made any kind of mistake on the approach it was not getting down sooner.) The only thing a non-precision approach (any approach without a glide slope) can do is get an airplane out of IFR conditions into VFR conditions close enough to the airport or runway so that a normal visual landing can be accomplished.

You are not allowed to go any lower than the MDA until visual, so get down to the MDA as soon as possible. The sooner you do go visual, more time is available to look for the airport, plan the pattern, and make a normal descent.

Legal Limitations

The legal limitations for circling are (1) keep the airport in sight at all times, except when losing sight "results only from a normal bank of the aircraft during the circling approach" and (2) do not leave the MDA unless you have visual reference for the intended

APPROACHES

runway and a descent to that runway can be made at a normal rate of descent using normal maneuvers.

In simple terms, if you lose sight of the airport execute a missed approach. During the Hilton Head approach the runway was not visible until on baseleg, but, because the circling was "tight," the airport was visible at all times. When the visibility is poor, as it usually is whenever a circling approach is required, stay in tight (which means speed must be down to minimum maneuvering speed and turns should have 25 to 30 degrees at bank) to avoid a mandatory missed approach situation.

In the first case the concern is, keeping the *airport* in sight while circling and in the second case the concern is with having visual reference to the *runway* before descent below MDA. What happens if the airport is visible but not the runway? That can happen when buildings on the airport, or maybe the tower or rotating beacon are seen, but the visibility is so poor that a pilot cannot see the runways, at least not well enough to know which is which.

Fly over and around the airport, if that helps, but do not drop below the MDA until positive the correct runway is in sight and a normal descent is possible. The minimum descent altitude will maintain clearance above the obstructions on the airport, but before descending below that, the FAA requests a clear shot at the runway, with no obstructions.

Certain pilots say do not descend below the MDA until within 30 degrees of the runway heading. The reasoning here is that any approach that is aligned within 30 degrees of the runway is considered to be a straight-in approach, so until you are within 30 degrees of the runway heading you can't descend any lower than circling minimums. This makes some sense, but there is no regulation that requires it. The only applicable regulation allows descent below the MDA anytime the runway or visual reference is in sight and a normal descent can be accomplished. This "rule" does make a very good guideline though: if possible wait until the airplane is aligned within 30 degrees of the runway before descending below the MDA.

Regulations are vague regarding what constitutes a normal rate of descent or a normal maneuver. Regulations are, however, very specific about what constitutes a visual reference for the intended runway. (Regulations might have one word like "normal," that can have endless interpretations and the next time you get a specific list.) There are 10 allowable visual references: approach light system, runway threshold, threshold markings, threshold lights, runway end identifier lights, VASI, touchdown zone, touchdown zone lights, runway or runway markings, and runway lights.

The FAA is trying to say that a pilot has to actually see the runway, or some part of the runway lighting system, to have "visual reference for the intended runway."

If the FAA had said this in the first place, there wouldn't have been any problem, but the reg used to say "the runway or the runway environment." The "runway environment" led to all kinds of problems because pilots were calling the highways and rivers and power lines around the airport the "runway environment," and pilots were using those references to find their way to the airport with predictably negative results. So the FAA "clarified" the meaning of runway environment with the list above.

Circling Approaches

Remember that "runway" means the actual runway and "visual reference" means the runway lights or markings, and a pilot must have one or more in sight to descend below the MDA.

Note that runway centerline lights are not included on that list. The reason is that if all a pilot saw was the centerline lights, he might think they were the runway lights and land to the side of them, instead of over them. With a small airplane, the runway might be wide enough to land between the centerline lights and edge of the runway, but a big airplane needs the whole runway. If a big airplane landed to one side of the centerline lights, it would probably end up with at least one set of wheels in the grass.

Fly legally and stay out of a lot of trouble: lose sight of the airport—execute missed approach; runway not in sight—do not leave the MDA.

Airspeed Control

What about flying a circling approach? What are the main things to keep in mind? One thing is speed control: not too fast (because that extends the circling radius) and not too slow (because that can result in a stall). Easy to say, but what does it mean? How do you determine the correct circling speed for a particular aircraft?

To determine the correct circling speed, start with the stalling speed of an airplane in the approach configuration, which probably means one notch of flaps, or 10–15 degrees, or the first increment-whatever the aircraft flight manual requires or the manufacturer recommends for approaches. The first flap position lowers the stalling speed and enables slower minimum speeds; further flap extensions add drag, which is useful for descent control, but not for speed reduction. We will call this speed V_{s1}

Having determined V_{s1}, multiply that speed by 1.3 to get final approach speed—a 30 percent margin over the stall, just like a short-field landing. (It won't be *exactly* the same number as a short-field landing because that number is based on 1.3 times the stall in the landing configuration, V_{so}.) Then add 10 knots to that number for maneuvering. This will take care of the fact that the stalling speed increases several knots while banking the aircraft during the circling maneuver.

This speed—1.3 times V_{s1} plus 10 knots— is the speed to fly while circling. Any faster than that will widen the turns much farther away from the runway. (This is why the circling minimums vary depending on the stalling speed of the airplane. The higher the stall, the greater the circling radius will be. Therefore, greater visibility is required to successfully execute the approach.) Any slower than 1.3 times V_{s1} plus 10 knots is too close to the stall.

Most pilots are so afraid of stalls that they err on the fast side, which would be okay, but the point is do not err. Know the circling speed, know how much power it takes to hold that speed in level flight, and practice until perfect. The goal should be to hold this speed within plus or minus five knots. That kind of control will yield a real edge on a circling approach.

APPROACHES

Altitude Control

Another important point is altitude control. This is always a big one on checkrides, and rightly so. The minimum altitudes that are allowed on circling approaches are quite low, usually much lower, in fact, than normal pattern altitudes. This puts an airplane quite close to the ground or anything sticking up from the ground. However, as long as the airplane stays within the protected radius for the circling maneuver and does not descend below the MDA obstruction clearance is assured.

The FAA could, of course, make the minimums higher, providing a greater margin of clearance, but then you would miss just that many more approaches (and possibly be just that much more tempted to "duck under" minimums.) So the FAA sets the minimum as low as safely possible. That means absolutely perfect altitude control—not one foot lower than the minimum circling altitude—until the runway is in sight.

The minimum altitudes on approaches mean exactly that: *minimum*. They are not assigned altitudes, which would imply a plus or minus 100 foot leeway, they are minimum altitudes: higher is okay but not lower. The problem with higher is that higher very often means up into the clouds, in which case a pilot is obligated to do an immediate missed approach.

The importance of accurate altitude control is obvious: not one foot lower than the minimum circling altitude (the MDA) until the runway or visual reference is in sight, and no higher than the MDA if higher is in the clouds.

If available, a flight director (FD) with a horizontal situation indicator (HSI) is an excellent reference to hold altitude on a circling approach. Set the heading bug to correspond to the heading necessary to fly that segment of the approach and punch on the altitude hold at the exact moment the minimum circling altitude is reached. Change the heading bug as the circling maneuver requires and fly the FD command bars. The command bars will immediately indicate minimum altitude variance, and the command bars are much easier for the eye to find after a glance outside to keep the airport and runway in sight.

It also helps to set up the HSI so the runway is represented by the OBS needle (FIG. 4-3). Point the OBS needle to the runway heading; if the runway is 32, put the head of the OBS on 320 degrees. (Don't worry about the middle part going full scale). The ends of the needle will represent the runway, the heading bug is the airplane's heading, and this will result in a "panel picture" of the circling maneuver. (Naturally, during a VOR approach, don't reset the OBS until after breakout and inbound approach guidance is unnecessary.)

Circling Patterns

This FD/HSI method depends on a pretty good idea what kind of circling pattern will be flown after breaking out, or you won't know the headings to position the heading bug. But even without an FD nor an HSI, it's important to anticipate the circle. Waiting until you break out to figure out how to get to the appropriate runway might mean trouble. A

Circling Approaches

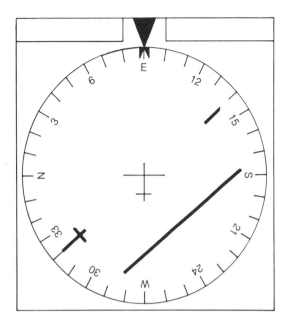

Fig. 4-3. *With the OBS set to the heading for the desired runway (top), the relationship between the aircraft heading and the runway will be shown on the HSI (bottom).*

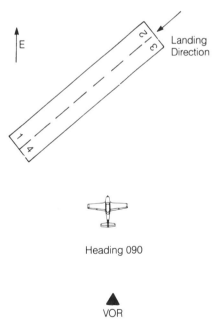

APPROACHES

plan that will vary with every circling approach is necessary. Let's break it down into some useful, general principles.

If at all possible, get in a position that is familiar, such as a normal traffic pattern. If you can get onto a downwind, fine. If not downwind easily, but base, that is almost as good.

The first choice is to get onto either a right or a left downwind leg, whichever is closer.

Landing from downwind is second nature, so you really don't have to think through the entire circling maneuver, only the part that gets to the downwind leg. Figure 4-4 shows several situations where a downwind pattern would be possible and the best way to maneuver to get downwind. (The approach into Hilton Head was a downwind pattern.)

The only part that will seem strange or unfamiliar about the circling downwind pattern will be the pattern altitude. Because circling minimums are usually lower than normal pattern altitudes the downwind leg will usually be much lower than normal. Hold the altitude until that point on base or final where that altitude *would* be normal, then go through

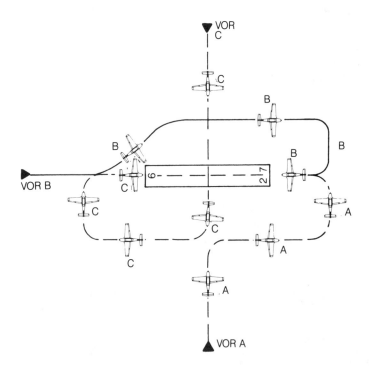

Fig. 4-4. *Three different situations where a downwind circling pattern would be appropriate: (1) A left downwind from VOR A to land on runway 27; (2) A right downwind from VOR B to land on runway 27; (3) An overhead, right downwind from VOR C to land on runway 9.*

Circling Approaches

the remaining items on the landing checklist—landing flaps, double-check the gear, propeller forward, and so on—and make a normal landing.

It is very common to make a circling approach onto a left or right base. The lower altitude won't be such a problem because you would normally be descending on the base leg anyway. You probably won't be lucky enough to set up exactly a base leg, usually a little maneuvering is necessary. Everything is back to normal on base, just like VFR. Figure 4-5 shows two examples of entries to base legs.

A tough one comes when breaking out directly over the touchdown point (FIG. 4-6). This situation is a little like the tennis shot that comes straight at your belt; you can't twist the racket around the right way to hit it. Likewise, in this situation a pilot can see the approach end of the runway underneath but can't seem to figure out how to get there.

One solution to this problem is shown in FIG. 4-7. It certainly means a lot of flying around, puts a new meaning into the word "circling," and if low weather or a big obstruction was nearby, use the pattern shown in FIG. 4-8. But the pattern in FIG. 4-7 has the enormous advantage of keeping the runway in sight at all times and it is all familiar: first a crosswind leg, then upwind, another crosswind, and finally downwind. The other way works and is sometimes necessary, but it involves figuring a 45-degree leg at a critical

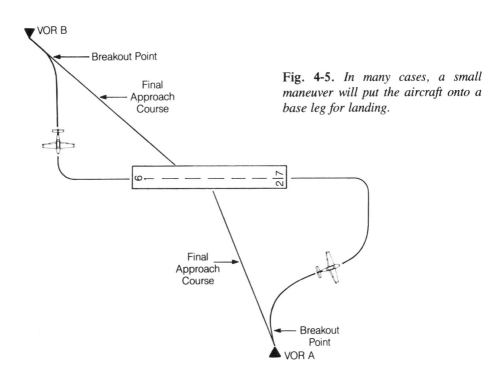

Fig. 4-5. *In many cases, a small maneuver will put the aircraft onto a base leg for landing.*

APPROACHES

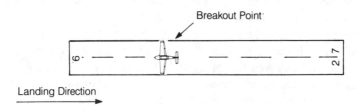

Fig. 4-6. *When the breakout point occurs over the desired touchdown point, but with the aircraft going in the opposite direction, a difficult circling situation is created.*

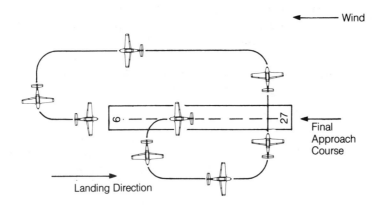

Fig. 4-7. *This is one solution to the problem of getting turned around for landing without losing sight of the airport.*

moment, and it means putting the runway behind the aircraft. (Remember, you can lose sight of the runway as long as the airport is visible.)

Circling approaches are tough and they are worth considering. When one comes up, stay cool, stay in control, expedite the final descent, plan ahead how to maneuver, watch the airspeed, and don't go below circling minimums until committed to landing.

NONPRECISION APPROACHES

Nonprecision approaches are sometimes called "letdowns" by older pilots. "Letdown" is probably a good name. The word "approach" gets used too many ways in aviation: "coming up on," as in "*approaching* the fix;" part of the traffic pattern, as in "make short *approach*" or "on final *approach*." Approach can refer to two entirely different instrument approaches: (1) a precision approach capable in its most advanced form of bringing the aircraft all the way down to the runway; (2) a nonprecision approach intended

Non-Precision Approaches

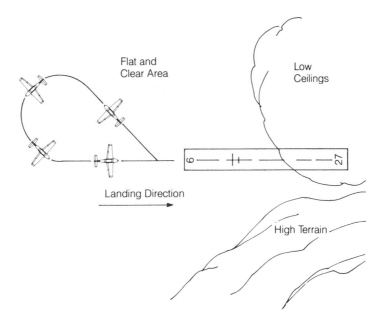

Fig. 4-8. *This pattern, which is very similar to a procedure turn, is another solution to the problem of getting turned around. It is the preferred pattern when high terrain or low ceilings preclude circling back and around.*

only to get the aircraft out of instrument conditions and into visual conditions in the vicinity of the airport. The term "letdown" for nonprecision approaches seems appropriate because that is all they really do: let an aircraft down, in a safe area, as close as possible to a runway or an airport. Calling VOR, NDB, LOC, and SDF approaches "letdowns" would eliminate one source of ambiguity with the word "approach."

Comment

I am probably not going to start any revolutions in terminology, but if we at least *think* of a nonprecision approach as a letdown, it clarifies what it can and cannot do, and what the pilot is trying to accomplish with it. A letdown is a controlled descent. Pilots control *where* the descending occurs with either a localizer course, a VOR radial, or a course from a radio beacon. Pilots control *when* the descending occurs with various fixes: course intercepts, VORs, nondirectional beacons, DME fixes, and crossing radials.

Initial Segment—Vectored

On certain nonprecision approaches, ATC vectors to the final approach course might be provided depending on whether radar is available or not. Normally, when radar is avail-

Approaches

able, an ILS is also available, and that is usually the preferred approach. But that isn't always the case. One of the more common approaches to JFK, for instance, is a VOR approach to Runway 13L.

Also, the smaller airports that lie under the approach control area of a large airport sometimes only have nonprecision approaches, but still have radar coverage. So, while a vectored nonprecision approach is not the norm, it is not unheard of either.

Whenever a vector is provided, an altitude will also be provided. Merely hold that altitude and wait for the inbound course to come in. When it does, turn inbound and continue the approach clearance.

Initial Segment—Nonvectored

Normally, nonprecision approaches will not be vectored. This means the pilot must establish the aircraft on the inbound course. This can be done either from any terminal routing (a radial from another VOR that intercepts the inbound course) marked "NoPt," which means "No Procedure turn," or with a procedure turn (a course reversal that intercepts the inbound course).

The minimum altitude for either a terminal routing or a procedure turn will be shown on the approach plate. When cleared for the approach, you may descend to the appropriate altitude for the terminal routing or procedure turn only when established on the routing or starting the procedure turn. This is the first step in the letdown procedure.

Many pilots think that the only reason for a procedure turn is getting turned around, which is true, but that's only part of the reason for a procedure turn. The main reason for a procedure turn is to get established on the inbound course without a vector or a terminal routing. To create an intercept, the only sure way is to start from a "known point," fly away from it, and then turn around far enough to create an intercept to the inbound course.

The "known point" to start from is the approach fix itself, either the VOR or the NDB, or, in the case of a localizer approach that requires a procedure turn, the beacon. The "flying away from it" part is outbound from the fix and the intercept part is the procedure turn inbound. Once established inbound, the airplane is in exactly the same position as if vectors had been provided.

Intermediate Segment

The intermediate segment of any approach is that part after the procedure turn vectors, or terminal routing, but before the final approach fix. The key to this segment is to stay as close as possible to the centerline of the approach course. For a VOR approach, use the approach mode on the flight director if available. If not available but with a heading bug on the directional gyro, use that bug to help bracket and track the course. Staying on the centerline of the approach course accomplishes two things: (1) remaining in the safest on-course area for the descent; (2) the centerline extends directly *to* and *over* the VOR or NDB.

Final Segment

It is very important to get as close to the VOR or NDB as possible. The only point in a nonprecision approach to accurately know where you are is when passing directly over the approach fix. Because the most important and critical descent in the approach comes after crossing the VOR or NDB inbound, it is vitally important that the descent be started from as close to the VOR or NDB as possible.

If the airplane passes by the final approach fix a half-mile off to the side it will be difficult to know when to start down. Station passage will be so slow that a pilot will not know with any accuracy when passage occurred and also will not know exactly where he is except off to the side of the desired course "somewhere." This is one way accidents happen—low, off course, in the clouds, groping.

The safest solution is to execute a missed approach and try to do a better job the next time. Or fly to the alternate, which—if picked carefully and conservatively—should have better weather or better approaches or both.

Just as most bad landings have origins with something done poorly earlier in the traffic pattern, most bad approaches usually have origins with something that wasn't properly executed in the initial stages of the approach: a procedure turn that wasn't carried out far enough, or a poor intercept of the inbound course, or an altitude that was allowed to slip. To keep things from getting out of hand, don't let them start. Remember that it is much easier to fix many small things than it is to fix one big one.

NDB APPROACHES

It is especially important to fly directly over the beacon on an NDB approach because it is the only way to add any real accuracy to what is the most nonprecise of all nonprecision approaches.

I'm going to tell you how most professional pilots fly NDB's, but don't tell them I told you because they don't like this stuff getting out. Do the best job possible to track inbound to the beacon, just like always. Be sure to go directly over the beacon, even if you have to "home" to it the last little bit. Look for a good, clean station passage—no slow, sliding by stuff.

Then, past the beacon, concentrate on flying the heading for the final approach course and don't worry about perfect tracking at this point. If there is no crosswind, the heading and the approach course will be the same anyway. If there *is* a crosswind, the airplane will drift slightly off the final approach course, but not very much and in any case it will still be parallel.

The key to an accurate NDB approach is starting the final segment from directly over the beacon—the final approach fix. The distance from the final approach fix to the missed approach point is usually no more than three or four miles. Over the beacon is already fairly close to the airport and the heading for the final approach will point directly at the airport. Tracking after beacon passage is nice but not essential—the airplane will not drift

APPROACHES

that much in three to five miles. It is also very easy to get it wrong, and turning the wrong way will take you much farther away from the correct course than any wind can. Here is a pop quiz to illustrate the point:

Question 1. The airplane has passed the beacon and is on the final approach inbound. The point of the automatic direction finder (ADF) needle is to the right of the tail and moving toward the bottom. In order to correct (a) turn right; (b) turn left; (c) wait awhile and see if it gets any worse; (d) experiment; (e) I'm not sure you need to correct—isn't the needle supposed to move toward the bottom?; (f) not enough information; (g) too much information; (h) may I execute a missed approach and go to the next question?

Question 2. (Something says one question is enough.)

Flying directly over the beacon and then flying the heading that corresponds to the final approach course should eliminate problems with interpretation, and any needle movement is caused by wind drift and nothing else.

An NDB is a letdown approach, pure and simple. Its purpose is to get down and out of the clouds in the vicinity of the airport. The NDB is no more than five miles from the airport. The final approach course heading will point directly at the airport and the clock will reveal fairly accurately when over the airport. That's all an NDB approach can do. Don't expect any more of it, and don't try to make it any more complicated than that. Flown properly, with careful attention to altitude, course intercept, station passage, heading, and time, if the weather is above minimums, the airport should be there.

ILS APPROACHES

An editorial several years ago in one of the general aviation magazines was complaining that the airlines were trying to hog all the federal airway money in order to have ILSs at all the airports served by the airlines. The editorial said the airline pilots considered any airport without an ILS to be inadequately equipped, and the editor disagreed. In fact, he implied that the airline types were a bunch of featherbedding sissies. I agreed with the editorial at the time. I figured it wouldn't hurt those guys to have to do an NDB or a VOR approach now and then.

Now I think I see what the airline types were getting at.

In a perfect world, all airports *should* have ILSs. An ILS is the safest landing approach available; any other approach is less safe—not *unsafe*, but *less* safe.

(Safety is not an absolute, it is a continuum that extends from wildly reckless to paralyzingly conservative. There is a large area in the middle that is "safe," but some parts of the middle are safer than others.)

An ILS approach is much safer than an NDB or VOR approach, and I put a much higher priority on ILS's now than I used to.

An ILS is also an integral part of the instrument system. (Why do you think they call it an "Instrument Landing System?") It isn't something tacked on to the end of the en route part of the airway system, and it isn't merely a method to get down out of the clouds. An ILS is a precision *approach* to the threshold of a runway (here the word "approach" is

ILS Approaches

exactly right). Instead of merely a letdown, hopefully into visual conditions, an ILS is a three-dimensional course aimed directly at the end of the runway.

If flying a VOR approach is like following an electronic highway to the airport, then an ILS approach is an electronic tunnel formed by the localizer course (for the sides) and the glideslope (for the top and bottom), and the light at the end of the tunnel is the approach end of the runway.

This chapter cannot rehash how to chase ILS needles down the localizer/glideslope. An instructor can help hone techniques for a particular airplane and instrument configuration, and after that it is just a matter of practice. However, remember a couple of important points.

An aircraft is seldom alone on an ILS. At any airport served by airline traffic, assume that there is an airplane three miles in front and another airplane three miles behind. Even at less busy airports assume that there are aircraft somewhere behind and in front. The only way to maintain separation is with speed control. If the controller requests 160 knots to the outer marker another airplane is bearing down at 160 knots. Do it, or if not, inform ATC for necessary traffic adjustments. Do not confirm 160 knots and then not do it.

If the ILS has a nondirectional beacon, and most do, this can be a tremendous help in staying oriented while being vectored around for the ILS, because the ADF needle always points to the beacon. Flying vectors for an ILS is like flying a big traffic pattern: downwind, base, and final (although sometimes entered on a base leg or straight-in) (FIG. 4-9).

On downwind vectors, the airplane is abeam the beacon when the ADF needle is 90 degrees off the nose. The farther downwind, the closer to the tail the needle will get. On

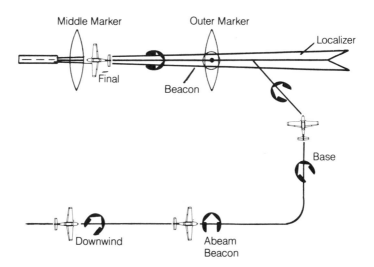

Fig. 4-9. *An ILS has a downwind, base, and final leg, just like a VFR traffic pattern. The ADF needle always points to the beacon, a useful aid in orientation.*

APPROACHES

base, the information is harder to interpret, but the ADF needle still points to the NDB, indicating at least a rough idea of position on the approach.

Naturally, the ADF needle points straight ahead when the airplane is established on the localizer inbound, then straight back after passage. Most beacons are colocated with the outer marker to become a good backup check on the marker.

ILSs are the bread-and-butter of the instrument flying system; they are accurate, reliable, and can safely handle large numbers of aircraft. It is hard to imagine the airway system without ILSs because they provide for a very nice flow from the higher speeds and altitudes of normal flight to the slower speeds and lower altitudes of the approach and landing.

The transitions are logical, smooth, and consistent, which is probably the key point the airlines were trying to make: an ILS approach is consistent. Every VOR approach is different, but, by and large, all ILSs are alike. The consistency and inherent accuracy of an ILS approach is what makes an otherwise critical maneuver—a blind descent to within 200 feet of the ground—a routine event.

PAR APPROACHES

A PAR (precision approach radar) is a precision approach, meaning it has a glideslope like an ILS, but unlike an ILS, a PAR is controlled from the ground, rather than from the air. PARs are not very common, but are still found at a few military or joint-use (combined military and civilian) airports. A PAR approach is a radar approach and it is the most accurate approach available—more accurate than an ILS.

The pilot's part on a PAR is easy: keep quiet and follow instructions. The controller will issue a missed approach procedure and instruct how to do the missed approach if communications are lost for a certain number of seconds followed by a request to discontinue acknowledging all further transmissions. ATC will then assign headings, within one degree of accuracy, to capture and track the "localizer," and will tell when to start the "glideslope" descent, and after that to increase or decrease the descent rate to track the glideslope.

The controller will be talking almost continuously so that positive communication is maintained; when he is not issuing a correction he will describe the position on the localizer and glideslope: "On the localizer, slightly above glideslope, correcting nicely." It is all very reassuring, easy to do and, most important of all, a PAR can lead right down to the centerline of the runway. The controller can even issue headings to make sure the rollout is on the runway. A PAR permits complete approaches when the fog is so thick that taxiing after landing is impossible and a "Follow Me" car is sent out to find you.

So what's the point of all this? If both the destination and alternate and every other airport in range goes below minimums—as they could if dense fog were to overspread an entire area—only two choices exist. One is to go to an airport with an ILS to 200 and $^1/_2$, and, using the emergency option, hope to shoot a perfect approach right down to the runway. The other choice is to go to a field with a PAR. Everything else being equal, the PAR is better.

An ILS *can* be used down to the deck and *is* used in a CAT III ILS, but a CAT III ILS requires the use of a flight director plus an autopilot system with autothrottles that have been certified for autolandings: a completely automatic landing including crosswind correction, flare, and centerline rollout.

Not only is the pilot not expected to be able to hand-fly the airplane to the ground, he is not even allowed to because the pilot's role in a CAT III ILS is to monitor the aircraft and be ready to take over and do a missed approach if anything goes wrong. Without this extra equipment and training, the PAR is simply the better approach when it comes to emergency approach conditions.

A PAR is *meant* to be hand-flown and it doesn't require any extra equipment or certification to be safe. (It might, however, require an emergency declaration to land.) It is a good idea to always have, filed away in the back of your head somewhere, the location of the nearest major military or joint-use airport: "Any safe port in a storm."

If you haven't ever performed a PAR, hire a knowledgeable instructor and practice a couple; the first one should not be the first one you *have* to do.

SAFETY VERSUS THE MISSION

I have not tried to teach how to fly instrument approaches in this chapter (that's a job for a flight instructor), nor have I dwelt on the basic approach terminology or procedures (See *The Aviator's Guide to Modern Navigation*, 2408 TAB Books.) These pointers gleaned from the real world of instrument flying should emphasize the importance of safety over mission accomplishment: "get homitis."

Approaches are critical because they are flown "low and slow," the most dangerous situation for an airplane. Approaches are high in workload and unforgiving of serious errors. Conservatism, practice, and attention to detail is the pilot's way of showing respect for the most demanding and the most rewarding part of instrument flying: the instrument approach.

5
Aircraft Limits

PILOTS TEND TO ACCEPT THE LIMITATIONS OF AN AIRPLANE WITH RELUCTANCE. They want the airplane to go anywhere, anytime, even though that isn't always possible. All airplanes have limits: a fighter, even though built to withstand stresses up to 9 Gs, can still come apart if pushed hard enough and freezing rain can keep any airplane on the ground, even an airliner. Power, strength, and weather avoidance systems extend capabilities, but all aircraft are limited to one degree or another.

Failure to know and observe whatever limitations exist is a fool's paradise. If a pilot elects to fly a single-engine airplane across an icy body of water without a raft, for instance, and does not contemplate the consequences of an engine failure under those circumstances, then the pilot is a happy fool: happy because he is unaware of the danger; foolish because if the engine quits, and it could, then he is almost certainly going to die from either exposure or drowning. Pilots that are unaware of the limitations of the aircraft are unknowingly taking these chances with themselves and with their passengers.

CONTINGENCIES AND BACKUPS

One of the keys to safe flying is accepting and dealing with unlikely contingencies: the unlikely "what ifs." "What if an engine quits right now? What will I do?" If there is no proper answer to that question, then you have discovered one limitation for the aircraft or type of operation. There are very often good answers to the questions, but the answers also very often involve some inconvenience or additional expense.

In the example above, when flying single-engine over icy water, an answer involving some inconvenience would be to fly around the body of water, while an answer involving

Aircraft Limits

an additional expense would be to invest in a raft. The wise pilot doesn't gamble when he can't afford to lose, and that means he always has a backup. If he doesn't have a backup, he doesn't do it.

Nonflyers seem to think that the only difference between a pilot's license and a driver's license is that the pilot's license is a little harder to get and a pilot can fly everywhere instead of driving. It just isn't true, of course. A car can be driven in almost any weather, assuming good equipment (snow tires, wipers, fog lights, etc.) and a safe driving speed, but the same cannot be said for the typical general aviation airplane (maybe not for any airplane). It doesn't take too long for even the newest student pilot to figure out that he cannot replace his car with an airplane. The advantage of extra speed is balanced by the disadvantage of limited utility.

This chapter is not intended to review a hard and fast list of limitations for particular aircraft types, but to start thinking about airplanes, their capabilities, and their limitations to develop a pretty good idea of what to expect from the airplane you do fly. Also, you might get a better idea of what money can do to make an airplane more useful—you probably have a pretty good idea already, but it never hurts to dream a little.

BASIC LIMITATIONS

The first airplane I tried to buy was about as basic as an airplane gets; an Aeronca Champion, otherwise known as an "Airknocker Champ." I was a second lieutenant stationed at Fort Knox, Kentucky, and a student pilot at the post flying club. There was a town just south of Fort Knox called Elizabethtown that had a little airport with one partially paved runway and a bunch of little airplanes that spent most of the time sitting in the sun. One of the airplanes was an Airknocker with a "For Sale" sign. It looked great—the fact that the fuselage was blue and the wings were white didn't bother me nearly as much as it should have—so I went ahead and made an appointment to meet the owners.

The owners turned out to be two old boys from E-town who made it pretty clear that they would be happy to take $1,900 for their treasure. Nineteen hundred dollars was an awful lot of money for a second lieutenant in 1968, so I sought and found, a partner.

(My increasingly incredulous new bride, who was just beginning to realize that she had not been given all the information she was entitled to prior to entering the marital contract, might have been a factor.)

Unfortunately, at the last minute my partner remembered that he was due to get out of the Army in a few months and I remembered that I was expecting orders to Vietnam, and somehow reason prevailed and we finally decided that maybe it wasn't such a great time to become aircraft owners.

A close call in some respects, but regrets still linger. It would have been a bag of worms (the reason the wings and fuselage were different colors, as it turned out, was because only the wings had been re-covered and the fuselage should have been), and I would have ended up having to buy out my partner, but I still wish I had bought it anyway. It may have been very limited but all I wanted and needed was a fun machine. If there were a few problems, so what? Solving those problems would have been part of the learning curve and maybe part of the fun.

Besides, I did not need to get somewhere. I was in the Army. Where was I going to go? I had no illusions about using it for transportation; if the Kentucky murk cleared to six or seven miles visibility, I could have built some time and confidence following roads and rivers for an hour or so.

If the engine quit, there were many fields to put it down in, and if I picked the wrong field, the Aeronea landed so slow I could hardly get hurt. I couldn't have gone far enough to get in trouble with the weather either, because range was only for about two hours and max cruise (downhill) was only about 80 mph; the farthest I could get away from home was only about 100 miles and still have some fuel. It would have been great. I might still buy one.

It is important to remember, that that airplane was originally designed and sold as *transportation*. When that airplane was built people thought there was going to be an airplane in every garage. It is easy to see now that the Champ had about as much utility as a glider. Keep that point in mind when considering what people expect of airplanes today.

ENGINE RELIABILITY

I did buy an airplane eventually, a "fully IFR" Cherokee Warrior: two coms, two navs, an ADF, full gyro panel, and pitot heat. It was a neat little airplane, and I picked up an instrument rating in it and flew in a lot of weather that I wouldn't think of flying in now (I did not fly in practically any kind of weather I wanted).

I considered the risks of flying single-engine IRF in marginal conditions and had plans for various contingencies: if the belt to the alternator broke I would tell ATC immediately, before the battery ran down; if the engine stopped in the clouds I would declare an emergency, descend on instruments (relying on a battery and maybe even a windmilling engine and therefore a generator) and when I broke out I would make an emergency landing in the nearest field; and I tried, as much as possible, to avoid mountainous and overwater routes. It was actually just wishful thinking. The simple and hard truth is that I assumed the engine would never quit.

And it didn't. (I knew all along it wouldn't.) I cheated the limitations of that aircraft for as long as I owned it, which wasn't too long. There is utility in an instrument-equipped, single-engine airplane with *real* contingency plans, but that doesn't mean a pilot can safely fly a single-engine airplane in any weather, or over any terrain.

One of the real problems with single-engine airplanes is that the engines *are* so incredibly reliable. Reliability does lead to complacency and a denial of the real risks. Pilots usually have many familiar explanations ready for nonflyers whenever the subject of single-engine reliability comes up;

Nonflyer: You fly an airplane with just one engine? What do you do when it quits?

Pilot: Aircraft engines don't quit. If your car were built and maintained and inspected and treated the way an airplane engine is, it would never break

AIRCRAFT LIMITS

down and would last 500,000 miles. Besides, an automobile has more parts than an airplane, like a water pump, clutch, or transmission. There aren't nearly as many things to go wrong on an airplane. And airplanes are inspected by highly trained and licensed mechanics, and then an additional preflight inspection is done by the pilot before every flight. The odds of anything slipping by are pretty remote.

Nonflyer: I know, but what happens if the engine does quit?

Pilot: Nothing happens. You just glide to a landing. It flies fine without the motor and can land in any field. There are plenty of places you can safely land it. Really.

Nonflyer: What about overwater, or in the mountains, or if flying in the clouds and can't see any fields, or it's night time? What then?

(This guy's a tough one, isn't he? A regular Sam Donaldson. Replay the tape:)

Pilot: "Aircraft engines don't quit. If your car...."

Let's take it from the top.

OVERWATER

A single-engine, *fixed-gear* aircraft should not normally be flown overwater out of gliding distance of land. A fixed-gear aircraft is very difficult to ditch successfully. The instant the wheels hit the water—no matter how gently they are set down—they stop, but the rest of the airplane continues forward. The airplane usually ends up flipping over onto its back. People usually survive, if all the belts are tight and shoulder harnesses are worn. Then comes the problem of getting out of a sinking airplane while hanging upside-down from a seatbelt. Flying single-engine, fixed-gear aircraft overwater, out of gliding range of land, is very risky business, even with a raft on board.

To stay within gliding distance of shore, get a good idea of the gliding range for any given altitude. As a general rule, most single-engine, fixed-gear aircraft have glide ratios of about 8:1—one mile of altitude (about 5,000 feet above ground level) will produce an eight-mile glide in a no-wind situation. If the pilot's operating handbook does not provide a gliding distance verses altitude chart, figure it out by climbing to 7,000 or 8,000 feet AGL (above ground level), accelerate to cruise speed, and then, over a known landmark, bring the power back to idle, decelerate to best glide speed and then descend 5,280 feet. (If the best glide speed is unknown, use the best rate-of-climb speed; this will be very close to best glide speed.) Don't forget the carb heat, if applicable, and clear the engine frequently during the descent. Note the location, measure the distance traveled in statute miles on the sectional, subtract one mile to compensate for the slight boost from the idling engine, and the result is a conservative idea of the glide ratio.

Assume a glide ratio of 8:1 and assume that the highest altitude filed is 11,000 feet (approximately two miles high); then the widest body of water to cross is 32 miles: 2 × 8 = 16 miles to the no-wind midpoint. That 32 miles equals Long Island Sound at the widest point, but none of the Great Lakes' widest points.

Overwater

Don't avoid water entirely though; climb to 11,000 feet and stay within 16 miles of the shore, a little closer to penetrate any headwind to reach the shore. To be truly safe, stay a little closer to have altitude "in your pocket" at the shore in order to set up an emergency traffic pattern over land.

This does not require much sacrifice. Perhaps climb higher than usual to cross a given body of water and perhaps go a little bit out of the way crossing any of the Great Lakes, but the inconvenience is minor compared to the peace of mind. (It probably does eliminate any trips to the Bahamas!) Remember, being an excellent swimmer or being in continuous radio contact with ATC isn't going to help much when trapped inside a semi-submerged airplane hanging upside down from a seatbelt. Fixed-gear aircraft cannot be reliably ditched—you might get lucky but do not count on it.

Retractable-gear aircraft *can* be ditched, but the success of the outcome varies in direct proportion to the calmness of the seas. Assuming an adequate life raft and life vests are on board and accessible, flying out of gliding range of the shore becomes a possibility, even for a single-engine airplane. Ditching is still not a pleasant thought, but at least it gives you a backup to answer to the "What if?" question: "If the engine quits, I shall ditch the airplane, gear-up and get in the raft."

Stay within gliding distance of the shore if any risk or risks seem insurmountable. (Most retractable gear, single-engine aircraft have glide ratios of approximately 10:1—a little better than fixed-gear aircraft.) No pilot likes to go out of his way, but the airplane is already going quite a bit faster than a car, which has to go all the way around a lake in any case, so going around is a fairly small price to pay to avoid possibly having to ditch.

Overwater routes present fewer limitations to a multiengine airplane. It is one reason people buy multiengine airplanes. If an engine quits, the remaining engine should get you to an airport—the second engine is the backup. It is still a good idea to have life vests, a floating ELT, and a raft with flares, lights, and other survival equipment for any flight more than 100 miles from land, and this is a regulation in certain instances.

There are still risks to overwater flying, even in a multiengine airplane. I flew from West Palm Beach to San Juan one moonless night in a Citation and found out the next day that a Beech Queen Air had gone down in the same area that same night. Rescuers found only an oil slick. Regardless of what happened, the airplane probably did not have a raft. Overwater flying does involve an additional element of risk, but the risks, with proper precautions, are acceptable in a multiengine aircraft.

Prior to a very long, maximum range overwater hop, check the single-engine range, which is usually shorter than the multiengine range; ensure that with engine loss at any point range will be sufficient to reach land, either by turning back or continuing.

(The point where the changeover occurs—where it is better to continue than to turn back—is called the "equal time point" (ETP). The ETP is not the same as the halfway point unless there is no wind, because the wind component is reversed. The formula for computing the ETP, along with a much more complete discussion of overwater navigation and flight planning, can be found in *The Aviator's Guide to Modern Navigation* 2408 TAB Books).

AIRCRAFT LIMITS

As a practical matter, as long as you avoid the maximum-effort overwater routes, you should be able to complete any overwater route that can be flown easily on two engines safely with one engine. There might be a short section in the middle where it is not obvious whether to turn around or go forward considering the winds, but if you have determined that you can make it at that point no matter which way you turn, you will have satisfied the most important requirement.

INHOSPITABLE TERRAIN

Inhospitable terrain is any area that lacks suitable emergency landing sites: wilderness, mountains, rough seas, heavily built-up urban areas. A single-engine aircraft should never be out of gliding distance of a suitable emergency landing site, which means single-engine aircraft should avoid inhospitable terrain. I used to fly in single-engine aircraft over mountainous terrain and over urban areas at low, but still legal, altitudes (1,000 feet), but not anymore. There aren't any "outs" to these situations. If the only engine quits over inhospitable terrain, the landing will be a crash. I am willing to accept some damage in an emergency landing and expect to be possibly "shaken up" a little—the potential emergency landing site doesn't have to be perfect—but I want to walk, or at least limp, away from any landing following an engine failure. Do not count on doing that in the mountains nor on city streets.

This does not eliminate flying over inhospitable terrain altogether. The gliding distance from altitudes of 10,000 or 11,000 feet for most single-engine aircraft is in the 16 to 20 statute mile range, which might well be enough to stay within gliding distance of an airport. (Look at *all* the airports, including private strips, to find that there are a lot of airports, even in the mountains) (FIG. 5-1). But it does limit the selection of routes in many cases. In the big, high mountain ranges like the Rockies or the Sierra Nevadas there might not be an acceptable route. A turbocharger and either oxygen or pressurization will enable a climb much higher than 11,000 feet, yielding a longer glide and making more routes available. As a general rule, the high mountains are no place for nonpressurized, single-engine airplanes.

Terrain is a consideration for multiengine airplanes at the higher altitudes where the inoperative engine ceiling is a factor. Single-engine ceilings for nonturbocharged twins are typically 6,000 to 8,000 feet. (Turbocharging will add several thousand feet to a single-engine ceiling.) This is fine for most of the country, but not for the high country in the West; pick routes very carefully out there. Flying several hundred miles out of the way is wiser than flying any route with a minimum obstruction clearance altitude (MOCA) higher than the single-engine service ceiling.

Trade airspeed for altitude with an inoperative engine by slowing to V_{yse} (best single-engine rate-of-climb speed) to have the slowest "drift down" and the highest single-engine altitude for more time and altitude to help clear the high terrain. But don't count on too much help. Assume a descent to the single-engine ceiling almost immediately and plan accordingly. Be careful to select a route that can be flown on one engine and terrain should not be a problem.

Urban Areas: Takeoff and Landing

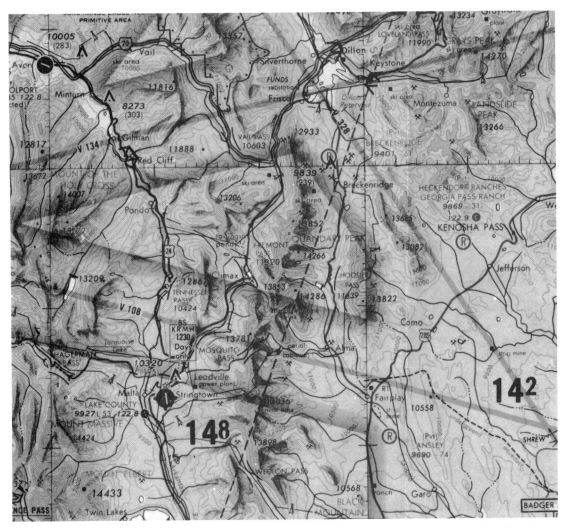

Fig. 5-1. *All airports, including private and restricted airports, are legitimate emergency landing sites.*

URBAN AREAS: TAKEOFF AND LANDING

Flying single-engine aircraft low and slow and out of gliding range of any suitable site is sometimes completely unavoidable over heavily built-up areas on takeoff and landing; regulations allow for this. One FAR, **Minimum safe altitudes; general.** says; *Except when necessary for takeoff and landing*, (italics added) no person may operate an aircraft below...an altitude allowing, if a power unit fails, an emergency landing without undue hazard to persons or property on the surface." In other words, operating under Part 91, the

AIRCRAFT LIMITS

FAA will allow a landing wherever an aircraft happens to come down, without penalty, when taking off or landing.

Flying over cities is not a problem at cruise altitude because there are always lots of airports within gliding distance around major urban areas. But it *is* a problem down low. The problem most often arises when pilots try to avoid flying over or through a terminal control area (TCA) going low, under the edges. It is always densely populated under a TCA, and flying low enough to get under the TCA usually means 3,500 feet AGL or lower, which is only about a five- or six-mile gliding range at the most.

Sneaking along single-engine under a TCA might be expedient, but not very smart and usually illegal because 3,500 feet or lower over an urban area is seldom "an altitude allowing, if a power unit fails, an emergency landing without undue hazard to persons or property on the surface."

If you insist on waiving the advantages of IFR and do go VFR on a route over a large urban area, go high and avoid the TCAs, except as necessary for takeoff and landing. (Mode C requirements have eliminated this option for many pilots anyway.)

SINGLE-ENGINE IFR

Single-engine IFR is tough. Using a single-engine aircraft for regular transportation means filing IFR and that inevitably means flying in the clouds. There is nothing wrong with flying single-engine aircraft in the clouds unless the clouds go all the way to the ground, in which case there is something wrong. If the clouds go all or even most of the way to the ground, a pilot is not going to find a safe place to land in the event of engine failure unless he happens to be very lucky.

Single-engine aircraft *can* be flown safely in the clouds, and should be operated IFR in any case, but one important limitation should be observed: fly only when the ceiling is high enough for a pilot to pick out and glide to an emergency landing site after descending through the ceiling following an engine failure.

Height of the ceiling required will vary with the terrain. In the Midwest, where the terrain is flat and fields are everywhere, 1,000 feet might be enough—enough altitude to fly a pattern. Over hilly, semicleared terrain such as parts of the South or Mid-Atlantic states, 3,000 or 4,000 feet; there are lots of fields, but there might not be one right underneath upon breakout. Over areas that are sparsely cleared and settled, like parts of Maine, Michigan, or Minnesota, 8,000 or 9,000 feet might be more like it. These are merely guidelines. Use personal judgment on this. Any area not suitable for an emergency landing is "inhospitable terrain" and should be avoided.

Perhaps it is obvious that this limits single-engine IFR to situations that are VFR anyway—at least 1,000 foot ceiling, and three or four miles visibility—but that still leaves a lot of utility. The ability to operate single-engine aircraft within the instrument environment and ATC system protects a pilot from the vagaries of scud-running and deteriorating or unexpected weather, and also provides a pilot with the same protection and assistance available to every other aircraft operating in the ATC system.

Single-Engine Night Flying

You don't *have* to observe this ceiling requirement (unless operating for hire). Part 91 does not specifically limit single-engine IFR to situations where sufficient VFR conditions exist beneath the clouds to make a safe emergency landing, but the regulation does not specifically allow single-engine IFR under these conditions either. If the engine quits, the burden of proof is on the pilot to show that the emergency landing was not unduly hazardous to persons or property on the surfaces and that the operation with ceilings lower than that necessary to find a suitable landing site was not careless and reckless. Think twice before you assume risks that the FAA has found unacceptable for paying passengers.

SINGLE-ENGINE NIGHT FLYING

Another major limitation of single-engine airplanes involves night flying. The inherent safety of single-engine aircraft is predicated on being able to make safe emergency landings. Darkness severely limits the ability to do this. The only time the ground can be seen adequately at night is when the sky is clear and the moon is full or almost full. (It also helps if the ground is covered with snow.) Otherwise, only lighted areas and dark areas are visible at night. That doesn't give much to go by. A dark area could be a field or woods; it could be flat or hilly, smooth or rocky. Whether a particular dark area is suitable as an emergency landing site is purely a matter of luck.

The only good situation to the limitations of single-engine night flying is to stay within gliding range of lighted airports at all times. It is usually possible to pick a route within gliding distance of a lighted airport at all times, zig-zag and fly higher if necessary. Naturally, the more altitude capability an aircraft has, the greater the gliding range and it is more likely that you will find a route within range of lighted airports. If having to zig-zag or fly higher is the difference between taking a chance on the engine not quitting that night and not taking a chance, that seems like a small price to pay. The need to stay within gliding range of a lighted airport at night is an unavoidable limitation of single-engine aircraft.

MULTIENGINE TAKEOFF PERFORMANCE

Weather, terrain, and time of day are major limiting factors affecting the utility of single-engine airplanes. The freedom to operate with low ceilings, over nearly any terrain, day or night, is the best reason for the additional expense and complexity of the multiengine airplane.

The next best reason is power. In practical terms it is very difficult to get more than about 400 horsepower out of an engine without going to a radial or turbine design; however, each has disadvantages: small turbines are very expensive while radials are heavy and bulky and have become practically obsolete. Therefore, if an aircraft requires more than 400 horsepower for adequate performance, the simplest and least expensive solution is to use two smaller reciprocating engines that equal the required horsepower. Two engines add to the expense of the airplane, but that is the only practical way to power airplanes

AIRCRAFT LIMITS

with loaded weights more than 4,500 pounds. This is also why multiengine airplanes do not fly so well on one engine—to a very large extent they have two engines because they need two engines.

The most serious limitation for multiengine aircraft is takeoff performance. For non-jet airplanes grossing less than 12,500 pounds there is no requirement to clear obstacles or even to climb to a specific altitude in the event of an engine failure after takeoff. The only requirement is that the aircraft be able to achieve a positive rate-of-climb, at sea level, in the takeoff configuration on one engine. Most piston twins can do a little better than a "positive" rate-of-climb (that could mean 20 feet per minute), and can eventually climb to 6,000 or 7,000 feet above sea level on one engine after being cleaned up. Being able to eventually climb to 6,000 or 7,000 feet is much different from having an assured takeoff and initial climb out capability.

Sudden loss of an engine on a multiengine airplane during the takeoff roll puts a handful of airplane on your hands. Instructors do not like to cut engines (mixture to idle cut off) on the takeoff roll unless the speed is very slow, because the resulting swerve when the engine does quit can be instantaneous and dramatic—sometimes too dramatic. In any case, at the higher speeds there is no way one little nosewheel can keep a multiengine airplane tracking straight down the runway with one engine pulling full-power torque and the other engine windmilling and creating a lot of drag, regardless of whether the power is lost suddenly or not.

The only way to keep it on the runway under these conditions is to reduce power on the good engine, and it has to be reduced *right now*. This is the reason "zero-zero" takeoffs are so dangerous in multiengine airplanes; without visual clues the airplane will almost certainly end up off the runway before the pilot can cut the good engine and regain control. Two engines do not make minimum visibility takeoffs (less than $1/4$ mile) safe: probably more dangerous.

If you have never had training in takeoff aborts, which this is, find a mature, experienced multiengine instructor and the widest, longest runway around, and practice. Start with an engine cut at the beginning of the takeoff roll, and work up to whatever speed the instructor feels can be handled, (*Do not* try this with a hood on. The lesson will be sufficiently impressive that there is no reason to take unnecessary chances.) If nothing else, you will learn firsthand the quick reaction necessary to keep the airplane on the runway, and the importance of aborted takeoff training.

When the airplane is airborne, but before the gear comes up, climb capability on one engine will be marginal to nil for most piston twins. If there is a lot of runway ahead, the best thing to do is to pull the power levers back and land straight ahead. (This applies only to Part 23 airplanes; Part 25 airplanes are committed to continue the takeoff at this point.) Without runway it is a real challenge to nurse the airplane along in the air while attempting to accelerate to at least V_{xse} (best single-engine angle-of-climb), then attempt to achieve a positive rate-of-climb so the gear can be retracted. One source of drag preventing the airplane from accelerating and then climbing is the gear hanging down, a real "Catch 22." If you've got the strength to hold full rudder as long as it takes, and if V_{xse} or better is maintained perfectly, and if the density altitude isn't too high, you should eventually be able to get the airplane to climb, which will enable you to retract the gear, but it takes a fine touch.

When, the gear is up, the *real* tricky part is over. (After that, everything is just tricky.) After the climb is established it should be possible to accelerate to V_{yse} (best single-engine rate-of-climb) and continue the climb-out and cleanup (flaps up, prop feathered, and engine secured according to the operating manual).

If you lose an engine on a multiengine airplane after the gear is up, a successful outcome will depend on pilot proficiency, the weather, and the terrain. If you can get the prop feathered without losing control, and if the airplane is not overloaded nor operating out of a very high airport, a climb should continue. If the terrain is flat and without obstructions, climb to a 400-foot pattern altitude and return to the field. If the weather is below pattern altitude, precluding an immediate return to the field, climb high enough to do an approach. If the airport is below landing minimums, continue the climb and go to the takeoff alternate.

TAKEOFF ALTERNATES

This is a good time to think about a takeoff alternate. Takeoff in a multiengine airplane from an airport where the weather is lower than 400 feet and/or below landing minimums practically demands that a pilot know where the nearest airport is that is above minimums in case there is a serious problem. Otherwise, what are the options? You can't circle back because the ceiling is too low and if the departure airport is below minimums an approach is impossible. You are going to have to go somewhere else that is above landing minimums and shoot an approach—that "somewhere else" is called a takeoff alternate.

A good guideline is that a takeoff alternate should be within 30 minutes of the departure airport at single-engine airspeeds. In the absence of an adequate takeoff alternate, the takeoff should be delayed until the ceiling rises to at least 400 feet for circling altitude.

ENGINE-OUT CLIMB GRADIENTS

If the terrain isn't flat, then the single-engine climb gradient—the ratio of altitude gained versus distance traveled—becomes the key factor. All multiengine piston-powered airplanes suffer from marginal single-engine climb characteristics. Some twins are better than others, but even the best will only climb about 400 feet per minute on one engine, and that only under ideal circumstances. Because the best single-engine rate-of-climb speed for most multiengine airplanes is around 120 mph (2 miles per minute), the best climb *gradient* to hope for is 200 feet per mile: two miles straight ahead to reach the minimum circling altitude (400 feet), five miles straight ahead to reach normal pattern altitude (1,000 feet), and 7-and-$^1/_2$ miles straight ahead to reach minimum initial approach altitude (1,500 feet).

This means that if hills or other obstructions are around the airport they must be visi-

AIRCRAFT LIMITS

ble to avoid them, because you almost certainly are unable to climb over them. One frightening thing I can imagine is sitting in an airplane with an engine out, in the clouds, climbing for all it is worth, wondering whether it will clear the hills or not. The only way to avoid that situation is to wait for the weather to lift enough to find a way through and around the hills.

The back of a Jeppesen approach plate sometimes has takeoff minimums for Part 135 and 121 operators (FIG. 5-2). These takeoff minimums allow for marginal single-engine climb capability and take into account the surrounding terrain. They are well worth observing, even though not required for Part 91 operators. If a SID (standard instrument departure) is available, it might describe required climb gradients, in feet per nautical mile.

Unfortunately, these restrictions are not in feet per minute (fpm), which is a more familiar way to measure climb capability. This is fairly easy to convert: if a multiengine airplane climbs on one engine at 100 knots (recall nautical miles per hour), divide that by 60 to get nautical miles per minute, 1.66 in this case. Then divide the expected single-engine climb rate by that number to get gradient in feet per mile. For a single-engine rate-of-climb of 250 fpm at 100 knots, divide 250 by 1.66 to get a gradient of 150 feet per mile. If the SID requires a climb gradient greater than that, do not go until the weather clears.

It would be great if the FAA published a "reverse approach" for each airport; the approach plate shows the safe way down and into the airport, perhaps the reverse approach plate (most likely it would be called a "departure plate") would show the safe way up and out for a given minimum rate-of-climb off a given runway. Certain pilots like to use an ILS "backwards" on climbout because what is safe going down must be safe going up. The only problem is that very few nonjet multiengine airplanes have enough power with an inoperative engine to climb a three-degree glideslope (more than 500 fpm at 100 knots). In the absence of an official instrument route around the obstructions, rely on visibility, and that means waiting until the ridges are not obscured.

When it comes to multiengine airplanes, for all the utility there is no guarantee they will *always* get you into the air and on your way. Multiengine airplanes do a pretty good job of maintaining level flight on one engine, but when it comes to takeoff and climb, they very often need both engines.

Comments

This is still a big improvement over the single-engine airplane. In a single-engine airplane, the entire flight is limited by the terrain and the ceiling; in a multiengine airplane, normally only the initial part of the flight is limited: takeoff roll, liftoff, and initial climbout. Ask the question, "If I lose an engine here, or here, or here, what will I do?" If the weather or the terrain, or a combination of both, precludes a satisfactory answer, then confront that limitation. All airplanes have limitations, no matter how much money is spent. If you want to operate airplanes safely you just have to live them.

Pilots tend to talk about how multiengine airplanes are more dangerous than single-engine airplanes: that the accident rates are much higher, that having two engines means

Engine-Out Climb Gradients

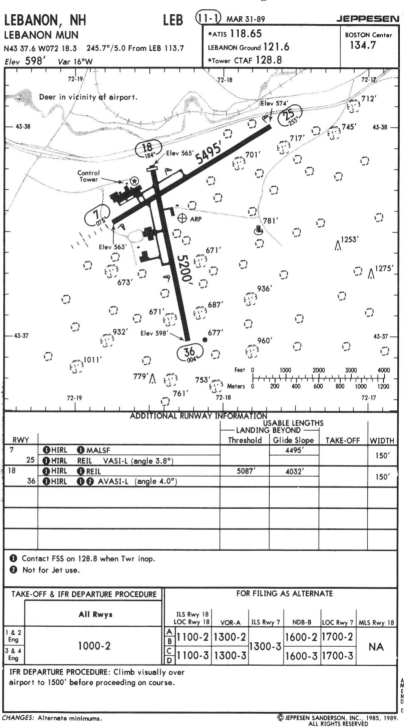

Fig. 5-2. *IFR takeoff and departure minimums and procedures are shown on the back of the first approach plate for any given airport. They are mandatory for Part 135 and 121 operators.*

AIRCRAFT LIMITS

two chances to lose an engine, that the remaining engine merely flies to the scene of the accident, and so on. I don't believe it. If true, multiengine airplanes would be equipped with a super quick, automatic shutdown device for the remaining engine in the event of engine failure—something like a bathroom circuit breaker that pops when there is a short circuit.

Then multiengine airplanes would be just as "safe" as single-engine airplanes because every time you lost any engine the other one would go with it and you would only have to worry about finding a place to put it, like a single-engine pilot. Multiengine airplanes are not so equipped, and never will be, because pilots want to have that engine available when needed and are willing to accept the risks inherent in engine-out flight if that is what is takes.

All this talk about multiengine airplanes being more dangerous than single-engine airplanes does point to a real problem: it is not that multiengine airplanes are not safe on one engine, but that they place a very high premium on pilot skill and proficiency to be safe, and multiengine airplanes are very unforgiving when that skill and proficiency is lacking.

Imagine that the pilot's requirement for skill and proficiency is a limitation of the airplane; "Multiengine aircraft are limited to being flown by well-trained and proficient pilots." If that limitation is not observed, then the multiengine airplane is more dangerous than the single-engine airplane.

TURBOPROPS

A turboprop is a turbine engine in which the exhaust gases drive a propeller. Turbine engines have many advantages over reciprocating engines, countered by several disadvantages. Turbine engines are small (relative to power), light (again, relative to power), easy to start, tolerate a wide range of operating temperatures, reliable, easy to maintain, and go a long time between overhauls. Unfortunately, turbine engines are expensive (even relative to power), not as fuel efficient as reciprocating engines (given current technology), and not as responsive as reciprocating engines (slower to "spool up" when power is applied). Despite these disadvantages, the advantages are so great that turbine engines of one kind or another are generally the engine of choice for aircraft requiring more than a total of 800 horsepower for adequate performance.

Aircraft with max gross takeoff weights between 8,000 and 12,500 pounds normally have turboprop engines. Aircraft grossing more than 12,500 pounds ("large aircraft," by FAA definition) normally have turbofan engines (exhaust gases drive fan blades). Due to improved fuel specifics relative to turbofans, turboprops are becoming more common on larger commuter aircraft. (The pure jet engine, or turbojet—no fan or propeller, only a compressor—is virtually obsolete, although it still has some limited military and supersonic applications.)

Most turboprops gross out at or under 12,500 and are certified under part 23, just like piston twins. Heavier aircraft, regardless of engine type, are normally certified under Part 25, which is a much more demanding set of certification standards. But there is no magic to a Part 25 airplane because performance is performance, no matter how an airplane is certified.

The main difference between a Part 25 aircraft and a Part 23 aircraft is the Part 25 aircraft has had its performance tested and measured in a way that assures a safe takeoff in the event of engine failure. Performance charts tell the pilot how much runway is necessary for a given ambient temperature, airport pressure altitude, takeoff weight, runway slope, and wind. The pilot then knows that he will be able to either stop the aircraft on the remaining runway or continue the takeoff (meaning, accelerate to rotation speed, lift off, achieve a positive rate-of-climb, retract the gear, raise the flaps, and continue the climbout with a climb gradient of at least 24 feet per 1000.) A Part 23 aircraft might well be able to successfully complete a takeoff through climbout also, but the information necessary to assure it in advance might not be available.

Regardless of how an airplane is certified, and to a large extent regardless of whether the turbine drives a propeller or a fan, turbine engines provide large amounts of power, and power enables a multiengine airplane to fly on one engine. If the pilot of a multiengine turboprop voluntarily observes the same limits required under Part 25, then he can have a virtually guaranteed takeoff. The power is usually there, but the takeoff limits have to be self-imposed. Because performance of the Part 23 airplane has not been documented, in most cases, as thoroughly as the Part 25 airplane, the limits have to be estimated.

Several variables affect takeoff distance for all airplanes: temperature, barometric pressure, wind, runway slope, and surface conditions. The only variable that can be easily controlled by the pilot is aircraft weight.

All the pilot of a Part 25 airplane does to ensure a safe takeoff is to match the takeoff weight to the conditions that exist at the time of takeoff: the lighter the airplane, less runway is needed to accelerate to rotation speed.

The pilot of a turboprop ensures that the aircraft can meet the accelerate-stop distance for the runway in use, even if it means reducing the fuel load.

Ideally, the pilot of a Part 23 turboprop would also like to have enough runway to *continue* the takeoff with an engine failure after liftoff, which means reducing the weight enough to meet the accelerate-go distance. ("Ideally," because in practice it might not be possible. A small turboprop is not a Part 25 aircraft and might not have the necessary power to do that.)

Only the Part 25 airplane has the information necessary to ensure a successful single-engine takeoff, and then only if the appropriate weight limitation is observed, however, a turboprop operated with a careful eye toward weight, still goes a long way toward eliminating most of the takeoff performance limitation of the piston twin.

JETS

The pilot of a Part 25 airplane, which basically means jets, has no choice but to observe the weight limitations dictated by the flight manual. If he does there will be enough runway to stop the airplane prior to V_1 and the airplane will fly and accelerate to a safe climb-out speed once past V_1, even if an engine quits. He still has to ensure terrain

AIRCRAFT LIMITS

and obstacle clearance because there is no guarantee a Part 25 airplane can outclimb the hills or jump over tall buildings, only that it can achieve a climb gradient of at least 2.4 percent. That translates into a climb rate of between 250 and 300 fpm for most jets, but most of the truly critical aspects of takeoff have been resolved.

The irony is that jets are easier than either piston or turbine twins to fly on one engine: set the correct pitch on the attitude indicator for single-engine climb, level the wings, and use the rudder for directional control: no prop to feather and little adverse yaw, at least compared to an aircraft with propeller-driven engines on the wing, there is very little yaw. Once you reach a safe maneuvering altitude, push the nose over a little to accelerate to flap retraction speed, run through the appropriate checklist, and fly away.

Recall from the Introduction that general aviation pilots have the toughest job in aviation, and this is one example. As easy as engine failures are to handle in jets, mistakes still happen, even with two- and three-man crews, which means that single pilots of propeller-driven, multiengine aircraft just can't be too careful.

WEATHER LIMITATIONS

The way an airplane is equipped for weather also determines its limits. Without radar (or other storm detection equipment), a pilot's ability to avoid thunderstorms is very limited, and without deicing (or anti-icing) equipment, true cold-weather instrument flying is virtually impossible. These are as much weather problems as they are aircraft problems and are covered in other chapters. (Remember that the airplane doesn't know anything about chapters.)

If the airplane won't fly because it is covered with ice, it really doesn't matter whether the pilot exceeded the limits of his aircraft, or the pilot showed poor judgment concerning the weather. The result is the same. All aircraft are limited in one way or another by the weather, and it is very important to remember and observe limitations.

COMMON SENSE

"Know and observe the limitations of your aircraft" is just another way of saying "Use common sense." If I had said that at the beginning of this chapter, it wouldn't have meant anything. I hope it is clear now. Common sense means do not assume the engine in a single-engine airplane will never quit. Common sense means do not expect 200 horsepower to drag an airplane through the air as well as 400 horsepower and certainly do not expect it to do a very good job if that 200 hp is stuck out on one wing, dragging a dead engine on the other wing along with it. Common sense means do not put yourself in a position where the answer is "I really don't know" when the question is "What are you going to do now?"

All airplanes have limitations, and being a good pilot means living with them. That is just common sense.

6
Personal Limits

"**K**NOW YOUR LIMITS AND DO NOT EXCEED THEM". I HAVE BEEN HEARING THAT SINCE I was a student pilot. It did not mean very much to me then and it doesn't mean very much to me now. It sounds good, but that is about all. It is easy to say things like that, and it is easy to nod a head in agreement, and there is probably no harm done in either. But merely saying it does not increase the overall safety level one iota.

Pilot limits are a direct function of the level of experience and training. There is nothing mysterious about limits and there is no reason to make limits any more difficult to understand than necessary. This chapter will attempt to provide a clear and specific idea of what is meant by "knowing your limits:" something concrete to use when faced with a situation, "Yes, I can handle this," or "No, I cannot handle this, not yet, but I am working on it, and I will be able to soon." I spent years trying to "know my limits" with virtually no guidance. Perhaps this chapter will guide you.

PRIVILEGES AND LIMITS

Student pilot limits are prescribed by the flight instructor. This recognizes that the student pilot does not have enough experience or training to determine limits.

A new private pilot is supposed to be able to determine his own limits. A private pilot's license has very few limitations—a low-time pilot *can* legally fly (in a Mode C-equipped airplane with two-way communications) from JFK to LAX, at night, with passengers, as long as he observes night VFR ceiling and visibility minimums. But that does not mean he *should* fly night VFR.

PERSONAL LIMITS

The privilege granted to a private pilot are enormous, which puts a tremendous premium on self-discipline.

INSTINCT

Too often pilots react to "know your limits" by relying on instinct or intuition—their "gut" reaction. They look at the sky, or the weather report, or the airplane, and they try to get a "feel" for whether going is the correct thing to do or not. The problem with this is that we are all cut out differently. Some people look at a new and potentially threatening situation—one they have no experience with—and just shrug their shoulders and off they go. Others are too cautious. They stick to the same routine: touch-and-goes, sightseeing, a short hop to get a hamburger at the one airport they are familiar with. They don't seem to be able to find enough confidence to try anything harder. Their "gut" reaction is an unreasonable fear.

Real anxiety about a trip—anxiety that seems to go beyond "butterflies"—do not ignore that feeling, which reflects taking on too much. Likewise, just because feelings of anxiety or fear are absent does not necessarily mean foolhardiness. But relying strictly on these "feelings" (or lack of them), prevents knowing real limitations. Real limitations can only be derived rationally.

A SYSTEMATIC APPROACH

When I was a kid the family moved around a lot because my father was in the Army. Moving all the time was normal because a different home very often meant living with other Army kids. Moving was normal for most friends, too. When you move all the time, you tend to develop a system for mapping out the territory and getting your bearings in a new place. Only retrospect has revealed this regular system for getting settled. The system was based on starting from a known point and expanding the base outward.

Specifically, a new kid went outside and hung around the yard. Sooner or later some kids would come over to check him out, and that would lead to going over to their house. And the next day the new kid would meet some other kids and so on. Within a week or so a kid knew the whole neighborhood, and within a couple of weeks also knew all the secret hiding places.

The process of knowing and observing limits as a pilot is exactly analogous to getting settled in a new area as a kid. A pilot starts from a known base—home airport—and extends his limits outward. But this base is more than a geographical base—it is also a base of skills, and it starts with those skills learned as a student pilot.

This is the opposite of what having a private pilot's license implies. The license allows almost anything, restricted by discretion and judgment. Perhaps *extending* with discretion and judgment would be better than *restricting*.

You have probably figured out by now that this problem of limitations is a problem only because the license permits so much freedom. Perhaps it would be much easier if the first license restricted you in some way, telling exactly how far away from home base you could fly, and what the weather had to be, and what kind of airplane you were allowed to fly, and so forth. (The new recreational pilot certificate does just that.) Then you wouldn't have to worry about limitations because you would always be bumping up against the restrictions of the license.

But each time you wanted to extend the privileges you would have to take a test to prove that you were capable of exceeding the limitation. That would be a real nuisance, it would be expensive, and it hasn't been the traditional method of private aviation in this country. But limits might get rid of that vague feeling in the pit of the stomach, standing on the ramp, wondering whether to go or not.

Goal-setting is always valuable, especially for a new private pilot, and is easier with knowledgeable guidance. This is an attempt to devise a rational plan for achieving those goals. Somewhere along the way every pilot should find a spot to come in.

NEW PRIVATE PILOTS

A brand new private pilot is truly limited to the specific skills learned for the private pilot checkride: short cross-countries in good weather, basic airwork, takeoffs, and landings. Those accomplishments don't come easily and any new private pilot has a right to be proud of them. Merely because he has a private pilot's license doesn't mean he knows everything necessary to safely exercise the full privileges of the "ticket."

Generally, the two best presents in the new private pilot's "bag" are takeoffs and landings. He has worked hard at them, done a lot of them, and enjoys doing them: his show pieces. He might not be able to do a soft- or short-field takeoff or landing very well, or might have a little trouble getting oriented in the traffic pattern at an unfamiliar airport, but he can do regular takeoffs and landings consistently well and safely.

The new private pilot can also do basic airwork fairly well: simple stalls, steep turns, ground reference maneuvers. He can plan a cross-country quite well, but is generally uncomfortable setting off on routes he either didn't cover as a student or that lack prominent landmarks like major rivers or four-lane highways. He can keep the airplane upright on instruments fairly well if he has to, and can usually manage turns to headings and controlled descents solely by reference to instruments. All in all, a pretty good start.

In practical terms the new private pilot has a lot of room to have fun, but very little capability for transportation. He can go sightseeing with a friend, he can do touch-and-goes, he can manage certain emergency situations, like approach to stalls and inadvertent instrument flight. But if a new pilot wants to go somewhere, he has to limit himself to the same kind of days and the same kind of trips prescribed by the flight instructor—namely, short day-trips out and back when the weather is good.

This is a good start, but what usually confuses the issue is that the first thing the typical new private pilot wants to do is to get out of the trainer and into something with a little more "performance" and extra seats because nobody has just one friend when he is a new

PERSONAL LIMITS

private pilot. So the first thing he usually does is get a checkout in the next bigger and more powerful airplane up from the trainer.

Unfortunately, "checkout" often means 30 minutes of dual, most of that in the pattern, and takeoffs and landings are what he does best and needs the least practice. The new private pilot gives up 40 or more hours of experience in one airplane for a mere half-hour familiarization in another. He gives up a good part of his limited competence because he wants to go just a little bit faster and carry two more people. It is hard to build on a known base when part of that base is given away.

The new pilot should get a lot more comfortable with the training airplane, particularly with the airwork while adding precision and smoothness to the flying in short forays.

But he is going to get a checkout in the bigger airplane, which is okay, but only if he is willing to spend five or six hours in it getting some dual instruction in all the maneuvers, just like the checkride. Then he will be as competent in the bigger airplane as in the trainer.

At this point the new private pilot can go sightseeing with his friends and make short cross-countries out and back. A new pilot actually wants to fly places that are far enough away to make the use of an airplane worthwhile, and probably does not want to have to come right straight back. In other words, he wants to use the airplane for real transportation. And that is when he starts running into trouble.

INSTRUMENT TRAINING

The only way to consistently, safely, and routinely use an airplane for transportation is to file IFR, which means the next step for a new private pilot is to get an instrument rating. Regardless of whether you are a new private pilot or have had a license awhile, if the goal is to use an airplane for regular transportation, and you don't have an instrument rating, start training.

Don't worry if you don't have 125 total hours (the minimum to take the instrument checkride). Go ahead and start anyway. The necessary hours will certainly accumulate, but even if they don't (by a few hours), you will still have the benefit of the instrument training during that entire time even if not actually filing IFR yet.

Upon completion of the instrument training you should have overcome any shyness about talking on the radio and probably will have discovered the secret for handling busy airports: IFR. When IFR, you are "in the system" to start with and while things might get busier near the "aerodrome," there won't be that sudden sense of being picked up and thrown into the middle of a maelstrom flying VFR into busy airspace. Two of the biggest problems for the new pilot—shyness with the radio and dealing with busy airports—are well on their way to being solved the day instrument training starts.

DAY TRIPS

In the meantime, before completing the instrument training, have fun and "extend" your capabilities safely with easy cross-countries. Pick a spot that interests family or friends to fly to for a short day-trip: a resort airport, an airport near a lake or the ocean, maybe an airport that has a specialty like gliders or antique aircraft.

Treat the trip strictly as fun but plan as carefully as if an instructor were looking over your shoulder. In fact, don't hesitate to ask an instructor to check you out or offer advice. (I have yet to meet an instructor who wasn't happy and even flattered to be asked to help a new pilot.)

If the weather is nice, go. If it rains that day, so what? You wouldn't want to go to a place like that on a lousy day anyway. Plan on staying only a short time. That way the weather won't have a chance to catch you. (But, of course, check the weather with flight service before heading home anyway and if it has turned marginal, stay until the weather clears, no matter how inconvenient or expensive.) This is the way a new private pilot can use his license in a safe way that recognizes his limitations, and gain a lot of confidence in the airplane and himself, which is the first step toward eliminating further limitations.

EXTENDED CROSS-COUNTRY FLYING

An instrument rating opens a new world—true cross-country flying. Until you have that rating there really isn't any point in even trying to use an airplane for reliable transportation, or consider going to busier airports, or flying at night, or trying to divine in some obscure way when the weather is "too bad" to go, or whether you can handle a cross-country requiring several stops—you shouldn't be doing any of these things anyway. They all involve a high probability that instrument flying skills will be required; without those skills and the rating that goes with them, you have no "good" choices. (Neither "scud-running" nor sitting in a motel waiting for the weather to clear are good choices).

Until you have an instrument rating, real cross-country flying exceeds your limitations. You may disagree, of course, especially if you live in an area of the country that generally has VFR weather. You are much more likely to agree with me after you have been embarrassed or you have scared yourself several times trying to prove me wrong. You can certainly become accomplished at flying in marginal VFR weather, but that's not the point. The point is that flying for transportation always involves certain elements outside your control, which includes but is not limited to weather, and the instrument flying system is the only reliable way to deal with those contingencies.

LIMITATIONS OF VISUAL FLIGHT RULES

The instrument flying system is also the *only* system because visual flight rules are not a "system" as such. For all practical purposes, VFR means "As long as you observe the hemispherical rule and stay out of the clouds, you can do just about anything you want, but watch out for yourself."

Visual flight rules do not provide separation, or any protection against deteriorating weather, and your access to assistance is several steps removed from merely picking up the mike and asking for help. A VFR pilot might have to look up an FSS or ATC frequency, try to raise them, possibly climb to establish contact, and then explain the situation.

PERSONAL LIMITS

Time and money are better spent obtaining an instrument rating than trying to get enough experience to extract marginal cross-country utility out of visual flight rules.

BENEFITS OF INSTRUMENT FLIGHT RULES

The immediate benefits of an instrument rating, beyond the obvious benefit of being able to fly in the clouds, are numerous. Filing IFR is one of the best things to make night flying safe. It is hard to get very far away from an airport on an instrument flight plan because the Victor airways tend to stay close to airports (many of the VORs are located on or near airports). In addition, you have immediate assistance available if needed because the controllers know their sectors extremely well, and if the engine does quit, they can steer you toward an airport, or away from unsuitable terrain. By planning routes to optimize the availability of lighted airports and by filing IFR you have gone a long way toward making single-engine night flying safe. Because you are operating on instruments anyway, the physical fact of flying at night is fairly irrelevant; the only real difference is that the windows are white in the daytime and black in the nighttime.

With the instrument rating, long stays are no longer a problem, nor are long trips. The main reason long stays or long trips are a problem for VFR pilots is that the weather changes over time and distance. If you stay anywhere long enough, the weather is sure to get worse, and if you fly far enough, you are bound to fly into some marginal weather. An instrument rating does not eliminate weather as a factor, but it does provide a very valuable tool for dealing with it.

Another big problem VFR pilots have that IFR pilots do not have is getting lost, or if not actually lost, the closest thing to it: not being able to find the destination airport. Instrument rated pilots certainly have to be careful when they fly to new areas—the VORs have unfamiliar names, the approaches have significant differences, and the airport layout might be confusing—but when all is said and done, an airway is an airway, an ILS is an ILS, and a runway is a runway.

The pilot on an instrument flight plan does not have to worry about getting lost over unfamiliar terrain, nor finding an airport, nor landing on the wrong runway; at least he does not have to worry nearly to the same extent the VFR pilot does. If an IFR pilot follows instrument procedures carefully, he virtually cannot help landing on the right runway at the right airport.

Comments

I once thought I had proved this to be wrong, though. I had a flight from Northamptom, Massachusetts, to Allentown, Pennsylvania, in a Piper Arrow. I had just received a flight instructor's license (including instrument instructor), so I had quite a bit of instrument training, but not a lot of instrument experience, and in particular I had never flown to Allentown before, nor had I ever had a flight that kept me in the clouds for the entire trip. (This was all done back when I trusted engines completely and didn't worry about flying single-engine IFR with low ceilings.)

Benefits of Instrument Flight Rules

The flight went uneventfully and routinely, I shot a nice (I thought) ILS approach to minimums, and I broke out at decision height with the runway right in front of me. So far so good. I made a normal landing, called ground control, and ground control cleared me to the ramp.

As I turned a corner and headed for the FBO, my heart almost stopped—the sign on the hangar said "Reading Aviation Service." Reading is just down the road from Allentown. I couldn't have shot an approach to the wrong airport because the frequencies would not have worked out right. (I had checked the ILS ident for the approach to Allentown and had received the correct ident.) The runway had the correct number and I was in radar contact the whole way—the controllers would not have let me make such a gross error.

But there it was, right in front of me: "Reading Aviation Service." I taxied to the ramp, completely confused, and tried to find something to confirm or deny that sign. I couldn't find anything, so I had to 'fess up to the lineman that I really wasn't sure where I was. He laughed and replied, "That sign gets a lot of people. You're at Allentown, all right. Reading Aviation has several locations and this is just one of them."

I was relieved, but not amused. It illustrates the point though: it is pretty hard to fly the correct approach to the wrong airport. The reason even experienced instrument-rated pilots still sometimes land at the wrong airport is because they cancel IFR and go visual too soon in marginal conditions. (I could tell you a story about that too, but I'm sworn to secrecy.) Landing at the wrong airport is a real faux pas. The "ultimate embarrassment" is the way a good pilot I know described it. Flying IFR and staying IFR is the best way to avoid it.

NEW INSTRUMENT-RATED PILOTS

Even though the VFR pilot solves a lot of problems the day he gets an instrument rating, limitations remain. Most limitations are related to inexperience with actual instrument conditions and not knowing how to operate smoothly within the system. If every new instrument-rated pilot flew awhile with different captains (experienced instrument pilots) he would very quickly get the experience necessary to fully and competently operate within the system. But he cannot, so he needs to know the limitations as a new instrument pilot to safely use the ticket and acquire the knowledge and experience necessary to extend his limits.

The new instrument pilot generally knows how to fly the airplane on instruments pretty well. He can follow the airways, and he has a pretty good idea of what it takes to shoot a good approach. In addition, after flying around with an instrument instructor for 40 plus hours, he is pretty comfortable with the airplane under normal circumstances.

He probably does not have much experience with flying in real clouds because most if not all his instrument experience will have been under the plastic cloud—the hood. In addition to not having much "actual" time, the brand-new instrument pilot has no solo experience flying instruments, and he probably hasn't had to deal with any emergencies nor abnormalities while on an instrument flight plan.

The new instrument pilot tends to be fairly well trained, but he lacks experience with

PERSONAL LIMITS

the reality of instrument flight. This is not his fault, but he should beware and acquire that experience in a safe way.

THE NEW CAPTAIN PRINCIPLE

Perhaps the best way to acquire the experience is to impose certain limits, in a very formal way, so a new instrument pilot can "get his feet wet" gradually and safely.

A new airline captain must increase personal approach minimums by 100 feet on the ceiling and $1/2$ mile on the visibility for the first 100 hours as PIC (50 hours if previously checked out in another type). In addition, the regulations for all commercial service, from air taxi on up to the major airlines, prohibit even *starting* an approach when the reported visibility is below minimums—there is only "look-see" capability is under Part 91.

This means when the new captain looks at the approach plate, he adds 100 feet to the decision height (DH) or minimum descent altitude (MDA), and $1/2$ mile to the required visibility for the approach. If the reported visibility is *below* the increased minimums, then he does not try the approach (he can't), and instead proceeds to the alternate. If the visibility is above minimums, he shoots the approach, but observes a 100 foot increase in the DH or MDA, even though the approach is perfectly safe for another 100 feet.

Part 91 does not have these restrictions, but if the restrictions are good enough for airline captains, surely they are good enough for new instrument pilots. A better idea would be to initially increase personal minimums by 200 and a mile. This will still offer many opportunities to complete instrument approaches; after all, increasing the minimums by 200 and a mile still permits a normal ILS to 400 feet and $1 1/2$ miles visibility, which is well below VFR minimums. This provides an extra 200 feet between the aircraft and the ground and the extra mile of visibility virtually assures finding the airport when you break out, even if you don't do a perfect job of keeping the needles centered (but any full-scale deflection still requires an immediate missed approach).

After 100 hours of instrument PIC time (not necessarily 100 hours of actual instrument time—that is a lot of actual instrument time—but 100 hours of flying on instrument flight plans) and at least 25 approaches, you can start thinking about decreasing the personal minimums to the new captain level of "plus 100 feet and $1/2$ mile" (300 feet and one mile for a standard ILS). After another 100 hours you should have enough confidence in your ability to shoot an accurate approach to published minimums. At some point you have to get out there in the real world and do some flying in actual weather, but do it safely and test yourself gradually. Increasing personal minimums is one way of doing that without an instructor watching.

ANXIETY

Probably the biggest problem for new instrument pilots is nervousness. Anything to eliminate anxiety will foster doing a better job of the task at hand. Fuel, for instance, is one thing you don't want to worry about. This doesn't mean "fill it up" all the time but it

Anxiety

does mean planning a good, fat reserve, and it means very careful and conservative flight planning with a complete flight log for every flight, even the short ones.

Freedom from nervousness also comes from having "solid gold" alternates: initially I think it is a good idea to increase personal minimums for an alternate to 1,000 and three (as opposed to 600 and two, the normal minimums for an alternate that has a full ILS.) It is awfully nice to know upon departure of the first solo, actual-instrument flight that if things go bad you can proceed to VFR conditions somewhere. You probably won't have to resort to it, but having a VFR alternate takes the pressure off knowing you have to complete an instrument approach prior to running out of gas.

Any pilot who is concerned about the anticipated approach segment of a flight, should plan and fuel the aircraft for a VFR alternate.

Also, don't worry about the aircraft. If there are any misgivings about the condition of the airplane, don't go. Misgivings will grow like monsters in the dark as soon as you enter the clouds. If the airplane has a history of even minor radio or gyro or engine problems—little ones like a gyro that is slow to spin up, or an engine that has started to burn slightly more oil, or a slight drop in the fuel pressure—and those problems cannot be resolved or satisfactorily explained, don't go. That gyro or engine or fuel pump will become a preoccupation as the flight goes on, and that is a set-up for making mistakes.

UPGRADING

After 40 hours of dual in an instrument trainer, you should be pretty comfortable with that airplane, so stick with it. Do not upgrade to a more sophisticated airplane until completely comfortable in the instrument flying system. Not only will the unfamiliarity of the more sophisticated airplane be a distraction from the job at hand, but its systems will require more attention and might malfunction. You have a lot to learn; take everything one step at a time.

The fresh instrument-rated pilot should learn how to handle the weather and learn how to deal with the instrument system. Get out there and learn how to flight plan, how to talk on the radio, how to fly the airways and shoot approaches, how to handle the little glitches that inevitably come up en route, and become comfortable with the airplane. Accomplish these goals, if you can, before encountering any "real" weather, even though the ability to operate in the clouds is ostensibly what an instrument ticket is all about.

An instrument rating isn't so much a license to fly in the clouds, as it is an admissions ticket to the ATC system. The ATC system is the important part, and that part can be learned when the weather is good. To do that, file IFR for every single flight, regardless of the weather. If you always file IFR, the actual weather part will take care of itself.

LOW CEILINGS AND VISIBILITIES

A multiengine airplane is necessary when the ceilings are low and visibilities are reduced. (See Chapter 5, Aircraft Limits.) This level of flying is reserved for experienced

Personal Limits

pilots who have already spent quite a bit of time learning first to fly on instruments, second getting completely comfortable with the ATC system, and third mastering at least one type of complex, single-engine airplane. To "put a pencil to it," this means 400 to 500 hours of flying time: at least 125 to get the rating, another 200 or so getting comfortable with the ATC system, and another 100 or so to be comfortable with a high-performance airplane. When that is done (however many hours it takes, the actual number of hours doesn't make any more difference here than it did when you soloed or received your first license), then you are ready to take on multiengine flying.

The two most important factors in flying a multiengine airplane IFR are pilot proficiency and the weather. Each factor is the subject of subsequent chapters—that is how important they are. But the process of knowing and observing limits as a new multiengine pilot is the same as new private pilots or new instrument pilots: start from a known base.

The "known base" is this case is skills obtained while training for the multiengine rating, specifically single-engine skills. The new multiengine pilot generally can do a pretty good job of handling the airplane on one engine; the hard part is maintaining that proficiency. There is no point trying to expand the base if you fail to maintain original proficiency.

Single-engine work gets all the attention in multiengine training, but it is also very important to maintain proficiency with the aircraft systems—electrical, hydraulic, gear, fuel, deice, and so on. This is a very important part of the "known base," because you cannot safely fly in weather that goes right down to the ground until you fully understand how an aircraft works. Flying in this kind of weather is going to be demanding enough; you don't need the distraction of systems incompletely mastered.

Mastering systems is what a checkride for a type-rating is all about. The examiner assumes that by the time a pilot gets to the point where he is ready to take a checkride for a type-rating, that the pilot knows how to fly instruments. (They sometimes find out differently, but that is the assumption.) An examiner expects to see an ability to handle aircraft systems, including abnormal and emergency situations, while operating within the ATC system. The PIC of a non-jet aircraft with a max gross less than 12,500 pounds does not have to take a checkride nor get a type-rating. However, because he is going to be the PIC of a complex multiengine aircraft, he should start from the same level of competence as if he were going to take a checkride for that airplane.

It is a good idea for new multiengine pilots to also raise their personal approach minimums by 200 feet and a mile for the first 100 hours, and then 100 and $1/2$ for another 100 hours. The extra margin of error takes a lot of the pressure off a "new multiengine captain." If the new multiengine pilot plans fuel loads carefully and conservatively and stays away from ice and thunderstorm, he should not have any trouble flying the airplane for its intended use: going places.

Flying a multiengine airplane in a variety of different and varying weather conditions is the ultimate extension of personal limits. When you reach this stage, you are basically doing the job of a professional pilot and success or failure will be determined by the same factors that determine his: proficiency, experience, maturity, and judgment.

CONCLUSION

All pilots have limits. What separates the capable pilot from the rest is an honest awareness of his limits. That awareness begins with a known base of technical competence upon which additional capabilities are built.

The point of this chapter was to help you find where on the continuum from student pilot to experienced, instrument-rated, multiengine pilot your "known base" is, and to provide concrete guidelines to extend your base beyond its present limits.

I hope I have succeeded, because the only alternative (other than further FAA regulation and control) is for you to scare the daylights out of yourself a bunch of times trying to figure it out by trial-and-error, and that is not a good way to do it, rest assured.

7
The Weather

YOU PROBABLY KNOW AS MUCH ABOUT THE WEATHER AS I DO. I'M NOT A WEATHERMAN, I am a pilot. I am not going to try to make you into a weatherman, either.

Weathermen—meteorologists—are scientists who have been trained to apply their knowledge, experience, and computers to the task of measuring, recording, and predicting the weather. They do a great job and I am happy to let them practice their craft in peace. But as Bob Dylan said, "You don't need a weatherman to know which way the wind blows."

Between the theory of forecasting and the reality of the elements sits the pilot. I cannot fly your trips, but I hope I can provide a way to deal with the weather and attempt to link the theory of weather forecasting with the reality of flight.

THEORY AND PRACTICE

Weathermen deal in generalizations and probabilities. I have yet to actually see weather that was as simple and straightforward as that shown on the weather charts: the lows do not have big Ls on them outdoors and the cold fronts are harder to find without the little pointed arrows. To understand the weather, and to describe it, requires pushing and pulling it into boxes with labels and lines and arrows. Weather charts are intentional simplifications, necessary generalizations. Charts reduce the weather to its lowest common denominator, which is helpful, but you won't always find the lowest common denominator on the other side of the windshield.

Pictures of real weather from space are great because a picture helps bridge the gap between the intentional simplifications of the charts and what is actually out there. The

The Weather

problem with satellite pictures, though, is that by the time you see them they are already history.

Records can be fed into a computer and the computer will explain what the weather did in the past, but the computer cannot guarantee that the weather will do the same thing in the future. Airplanes can be built to withstand gust factors of 50 feet per second and carry or shed large amounts of ice, but the weather will not always agree to limit itself to gusts of 50 fps or large amounts of ice.

Real weather that is actually out there—not the stuff on the charts nor in the computer nor on yesterday's satellite photo—won't stand still to organize into neat highs and lows with fronts that start and stop right where the line is. The real-time weather and the weather on the charts is not always the same thing.

Nonetheless, a pilot should know about the weather from a theoretical point of view. Without an understanding of the theory behind the forecasts, there can be no basis for dealing with the weather that does exist. A good place to start to learn about the theory of weather is with *Aviation Weather*, an FAA publication. This book is particularly good at explaining and illustrating the basic concepts of weather theory.

If meteorology seems a little tough going, remember, weather is applied physics; there is no way to make it easy and get beyond, "High pressure means good weather, and low pressure means bad weather." It is worth the effort though. Bob Buck's *Weather Flying* (Macmillan, 1970), is a classic on the subject, and is also well worth checking out.

There are, of course, numerous others, and I cannot think of any that are not valuable, each in its own way. Each has a little different angle and that is what you need to deal safely with the weather—the angles—because nobody has all the answers and the weather itself is not going to make it easy.

Perhaps there is a best angle—leave the weather to the weathermen. Learn everything possible about the weather, but do not try to outguess the guys who do it for a living. The experts make mistakes often enough; how well do you expect to do? *Optimize the meteorologists' abilities to minimize personal risks.*

I've been playing this game of amateur weatherman for years, trying to predict the weather from the charts and maps independently from the weather forecasters. I know all about jet streams and upper level troughs and steepness of pressure gradients, and still miss all the time. Sometimes throwing darts might be better. Other times I think a trained monkey could do better than me. In fact, it seems like I did better when reading a barometer: if it said "Change," the weather usually changed.

(I am thinking about getting one of those little houses with the people that come out with umbrellas if it is going to rain. I am going to set myself up as a consultant. I will have an 800 number, and pilots will call me from all over the world, and if the little people are out there with their umbrellas, I will say, "Do not go. Looks like it is gonna' rain." I am going to focus advertising on all the people who do not believe in weathermen. I am gonna' make a million bucks.)

TIME ELEMENT

Weather will not stand still; it is slightly different this moment from the moment before and will change yet again in the next moment. On a moment by moment basis people usually are not especially aware of the changes, but the changes are taking place just the same. Anytime you are dealing with any aspect of the weather it is important to always ask "What is the time factor?"

It is very important to have it straight whether you are looking at past history, a present fact, or a future expectation. It is too easy to get a weather briefing and forget that part of the briefing is history, part of it is current, and the rest is speculation.

Past weather provides an idea why the present is what it is, and that is helpful, especially to the forecaster. But it does not assist the planning process—the weather strategy. That strategy has to come from present, short-term fact (mainly the current sequence reports), and future, longer-term speculation (mainly the terminal and area forecasts).

FORECASTING

"Leave the weather to the weathermen" means leave the weather forecasting to the professionals. Bob Buck's book on weather is a classic for good reason, but Captain Buck and I might disagree a little on this particular issue. The way I read *Weather Flying* (and most books on weather and aviation), Buck seems to say that if you read enough books on the weather, and if you fly in enough weather, that eventually you will start to have enough knowledge and experience to "read between the lines" of the forecasts and fine-tune and improve upon the official forecasts. Maybe Bob Buck can. Maybe he can look at a forecast and a surface chart and put the two together into something that is better than the two parts, but I cannot with any consistency. Frankly, I don't think any pilot can.

Everyone plays the game, standing there looking at all the charts: "I know the forecast says it is going to be 400 indefinite, $^3/_4$-mile visibility with light rain and fog, but I think that low is going to push on out of here tonight...400 and $^3/_4$ is still above minimums at any rate, but I don't think it is going to be that bad." And often the pilot is right. The problem is those are the only times a pilot remembers: incorrect personal forecasts are forgotten, fostering a distorted view of forecast abilities.

Forecasting the weather is and always will be a matter of playing the odds; nobody is right all the time. There is nothing wrong with playing the game and comparing results, and there certainly is nothing wrong with learning all you can about the theory behind the forecasts, especially if that knowledge is used to make the official forecasts more conservative ("I do not think that low is going to move as fast as they think. I think I shall add 30 minutes more fuel for weather delays."), but if you cannot be choosy, go with a winner, and the guys with the best record are the professional meteorologists. Use what they offer to optimize your ability to deal with the weather.

THE WEATHER

THE BRIEFING

The key element in dealing with the weather is the weather briefing because the weather briefing is the link between meteorologist and pilot. Specifically, the weather briefing forms the basis for much of the actual flight planning, for the identification of alternates, and for the avoidance of hazardous weather. Getting a good weather briefing is not terribly difficult, but it is not automatic either. A pilot cannot merely blow into a flight service station (FSS), or make a quick phone call, and go. He has to work at it a little.

When it comes to weather briefings, professional pilots are no different than anyone else: what they say and what they do are very often two different things. Ask most professional pilots how they deal with the weather prior to a flight and they will tell you that they always try to go to a weather briefing facility (as opposed to making a phone call), and that they study, at length and in detail, the surface progs and the synopses and 300 and 500 mb charts and everything else hanging on the wall before taking a look at the sequence reports and forecasts.

A lot of professional pilots brief themselves on the weather, and the first thing they usually look at is the current weather, and the second is the forecast, and the third is the winds aloft, and after they have that information for as many stations as they can think of, they might glance at the charts on the walls on the way out.

In any case, as FSSs become harder and harder to find, most of the time pilots merely make a phone call and ask for the destination sequence and forecast and the winds aloft along the route of flight, and then check out an alternate.

That is not the way to do it, but there are several reasons why they get away with it:

1. They are going to file IFR in any case, so they don't need to look at the VFR possibilities along the entire route of flight.
2. They are probably flying an airplane with two or more engines, a complete flight director/autopilot system, full deice or anti-ice capability, and top-of-the-line radar, so they are not too worried about a lot of the weather in the first place.
3. They are very alert to weather that all pilots worry about: low ceilings and fog, moderate to severe icing and/or turbulence, freezing rain, snowstorms, severe thunderstorms, squall lines.
4. They learned long ago to always have an alternate, so they don't spend a lot of time figuring out whether they need one or not, all they have to do is pick one.

The only thing wrong with this practice is that you should not try to brief yourself on the weather no matter who you are or how much experience you have. If you want a complete weather briefing, ask for it. (I know it is a long shot, but it is so crazy, it just might work.) Both National Weather Service and flight service station personnel are trained to give weather briefings. They know the format, they know where to get the information, they usually know what is relevant and what isn't, and in person they can provide a hard copy.

Pilots know this, but egos might get in the way and they brief themselves on the weather. Very few briefers will fight them; the more work the pilot does, the less work a briefer has to do. But briefers can and should do the briefing. That is their job. Self-briefers are dumb. It is just too easy to miss important information.

Specifically requesting a complete weather briefing also forces the briefer to concentrate on the facts. Remember, a flight service specialist is not a meteorologist. He is a trained weather briefer, not trained nor qualified to analyze or forecast the weather; nor are you. You want to know what qualified meteorologists say about the weather along the route of flight and from that information you expect to make certain decisions concerning that flight. You are not interested in the briefers judgment—except concerning the advisability of VFR flight (more on this later)—nor do you want him to make decisions for you: not his job, yours.

The military has an excellent system for handling weather briefings: a pilot *will* receive a standard weather briefing, he *will* receive a hard copy of the briefing, the pilot and the weather briefer *will* sign off on it to show that it has been accomplished, and a copy *will* be retained in case there is a need to refer back to it. In the military, a pilot must get a complete weather briefing; if he does not have a signed copy of the briefing, the aircraft is not dispatched. The military can be hardnosed, but the system works, and every pilot would be wise to copy it to the extent that is possible.

FAA STANDARD WEATHER BRIEFING

The FAA calls a complete weather briefing a "standard weather briefing." Request a complete weather briefing and the briefer will understand, in most cases, that you want what he calls a "standard briefing," but if you know the jargon, that eliminates any possible misunderstanding or confusion.

Standard briefings, as described in the *Airman's Information Manual*, have 10 parts:

1. Adverse conditions.
2. VFR flight recommendation.
3. Synopsis.
4. Current conditions.
5. En route forecast.
6. Destination forecast.
7. Winds aloft.
8. Notices to Airmen (NOTAMs).
9. ATC delays.
10. Other information available on request.

There is a reason the standard briefing is organized this way and there is a reason for each part.

Adverse Conditions

Adverse conditions are conditions that might lead to a cancellation or modification of the proposed flight: icing, turbulence, thunderstorms, IMC (instrument meteorological conditions) for VFR flights, runway or airport closures, any adverse condition possibly affecting the safe outcome of the flight as proposed. The reason this information is given first is fairly obvious: there is no point in proceeding with the rest of the briefing until these adverse conditions have been taken into account. If the conditions are sufficiently adverse, the entire flight might have to be canceled, in which case there is no need to continue with the briefing. In many cases the flight will not be canceled, but changes will have to be made—IFR instead of VFR, changing routes to avoid an area of thunderstorms—and the briefing will then have to be changed. So it makes sense to look at adverse conditions first.

VFR Flight Recommendations

Whenever a pilot proposes to go VFR and, in the briefer's judgment, VFR flight would not be advisable, the briefer will advise the pilot that "VFR flight is not recommended." This does not mean that the pilot cannot go VFR if, in *his* judgment, VFR flight *is* safe. The final decision is entirely the pilot's responsibility, but it is a recommendation that should be taken very seriously. Again, it comes early in the briefing for the simple reason that the rest of the briefing might not be necessary if the pilot elects to follow the briefer's recommendation and cancel the flight. Alternately, if the pilot is instrument rated and current and has access to an instrument-equipped airplane, he might elect to go IFR instead, which also needs to be taken into account by the weather briefer during the rest of the briefing.

Synopsis

This is the big picture. The briefer will attempt to describe (assuming a telephone briefing) the major weather systems affecting the route of flight and the systems' general movement. In person the briefer will use wall charts and diagrams to show areas of high and low pressure, frontal movements, wind flow, and areas of marginal VFR and IFR weather. (FIG. 7-1) This is all "good-to-know" stuff, and will foster an understanding of the reasons for the actual and forecast weather in the briefing. It is also good background information for understanding later during the flight what is happening when things do not go exactly as forecast or reported.

Current Conditions

The content of this part of the briefing is fairly obvious to most pilots. (FIG. 7-2) The briefer will provide current conditions for the departure airport, selected en route airports, and, of course, for the destination airport. He might modify the en route portion depending on the type of aircraft and the flight rules under which the flight is proposed: a pilot

FAA Standard Weather Briefing

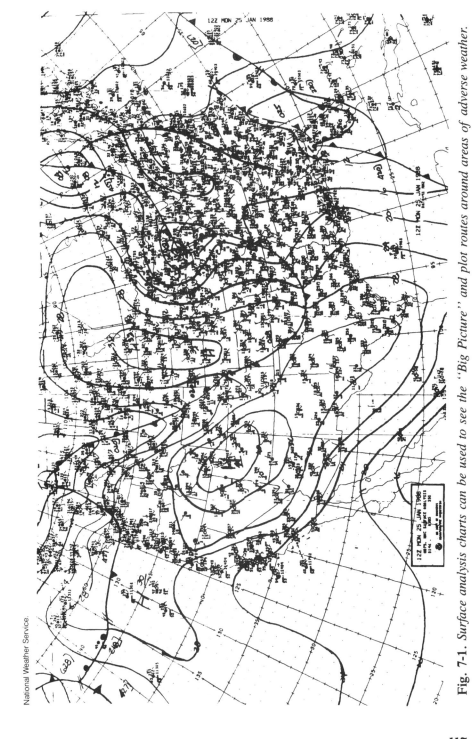

National Weather Service.

Fig. 7-1. Surface analysis charts can be used to see the "Big Picture" and plot routes around areas of adverse weather.

The Weather

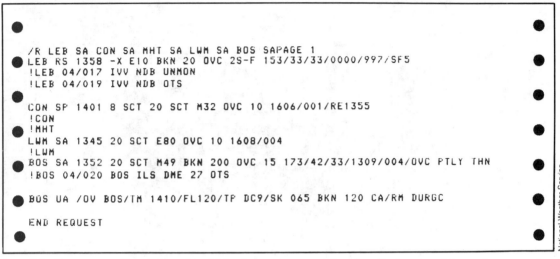

Fig. 7-2. *Sequence reports, abbreviated SA, show current weather for all airports with weather reporting service. Notices to Airman (NOTAMs) for individual airports are included with the appropriate sequence report, preceded by an exclamation point. A pilot report is indicated by the abbreviation UA.*

proposing a flight under IFR in a multiengine aircraft is much more concerned with the current conditions at the destination than he is with the en route conditions; the pilot of a single-engine aircraft is very concerned with current en route conditions regardless of whether he is VFR or IFR—he needs to have sufficient visibility along the entire route of flight to be able to find an emergency landing site if the engine fails. (FIG. 7-3) The VFR-only pilot, single- or multiengine, needs to know the current conditions for as many airports as possible along the entire route of flight in order to simply maintain VFR at all times.

En Route Forecast

The en route forecast portion of the weather briefing is also fairly self-explanatory. (FIG. 7-4) Again, as with the current en route conditions, the en route forecast is of much greater importance to the VFR-only pilot of a single-engine aircraft than it is to the instrument rated pilot of a multiengine aircraft. The ability to fly "in the weather" en route is one of the main reasons for filing IFR, and the ability to do that without too much regard for the ceilings en route is one of the main reasons pilots operate multiengine aircraft: the second engine takes most of the sweat out of emergency landings.

The pilot proposing a VFR flight in a single-engine aircraft is very interested in the en route forecast because he must stay VFR at all times along the route. He needs to know, in some detail, what to expect for weather from takeoff to cruise to descent. He cannot afford to have any gaps en route because he can't merely plow through the cumulus and fog and haze and rain showers the way the pilot on the instrument flight plan can, and he needs to be able to see and find suitable emergency landing sites along the entire route of flight.

FAA Standard Weather Briefing

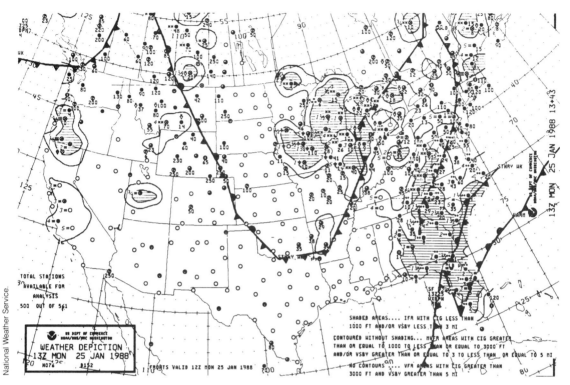

Fig. 7-3. *Weather depiction charts present the current weather conditions. Shaded areas represent current IFR conditions. Contoured areas without shading are experiencing marginal VFR; other areas are good VFR.*

The pilot of a single-engine aircraft on an instrument flight plan can plow through most of the weather, but he still needs to find a suitable emergency landing site. His concern during this portion of the briefing is ceilings and visibilities en route. Specifically, he needs to have enough time and enough in-flight visibility after he breaks out during an emergency descent to locate smooth terrain. Over flat terrain with lots of fields he only needs a thousand feet or so and three or four miles visibility. Over hilly, wooded terrain he would want higher ceilings and better visibilities. The en route forecast portion of the briefing is the information needed to ensure being able to make those emergency landings en route.

The pilot of a multiengine aircraft on an instrument flight plan is not completely disinterested in the enroute forecast. He still must be prepared for possible precautionary landings resulting from fuel or maintenance problems, or adverse weather. The en route forecast is not nearly so critical for him as it is for the VFR pilot, or even for the pilot operating single-engine IFR. The multiengine IFR pilot needs to have at least an idea of what the forecast is for various points along the route of flight. Freedom from most of the en route weather worries is where all that money spent for the second engine pays off.

THE WEATHER

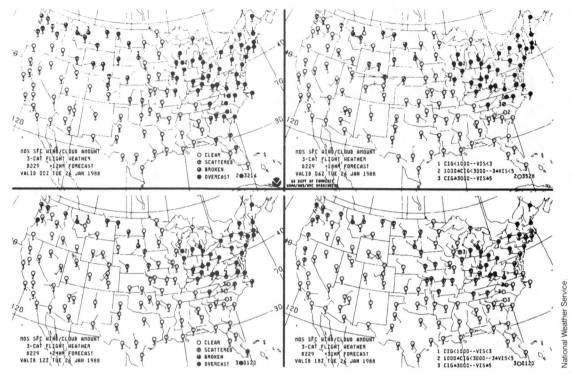

Fig. 7-4. *Aviation weather forecasts for 12, 18, 24, and 30 hours are shown on this set of three-category flight weather charts. Station symbols depict surface wind speed and direction, cloud cover, and IFR, marginal VFR, or VFR conditions.*

Destination Forecast

The destination forecast is just one of a string of forecasts that the VFR-only pilot needs. And not only whether the destination is expected to be VFR upon arrival, but that all of the en route stations are also expected to be VFR. (FIG. 7-5) The destination forecast as part of the standard weather briefing is merely a continuation of the en route forecast.

An IFR pilot regards the destination forecast as one of the most important parts of the weather briefing. Zero visibility is permitted en route, but visual reference is required to land. The destination forecast explains what weather is expected before, during, and at the planned arrival time, and therefore explains whether he can expect to shoot an instrument approach or not, and if so, to what minimums, and whether or not an alternate airport will be required. This is important flight planning information, especially applied to fuel planning, but the destination forecast is critical to all aspects of flight planning and needs to be understood completely and considered thoroughly.

FAA Standard Weather Briefing

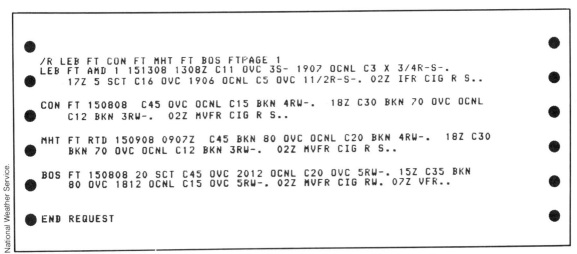

Fig. 7-5. *Destination forecasts are often broken down into several different segments for greater accuracy. The last segment is always a general outlook forecast. Qualifiers such as OCNL (occasional) are among the most important parts of these forecasts, and must be taken into account during flight planning.*

Winds Aloft

The winds aloft are *forecasts* of expected wind direction, velocity, and temperature for altitudes beginning at 3,000 feet AGL (above ground level) and continuing at 3,000-foot intervals through 12,000 feet MSL (mean sea level). (FIG. 7-6) For altitudes above 12,000 feet, the sequence continues at 6,000-foot intervals through 30,000 feet MSL and above that the winds aloft are reported for 34,000 and 39,000 feet, for reasons only the National Weather Service understands. (Winds aloft are also available on special request for altitudes above that. When you finally breakdown and get that SST you have always wanted, or perhaps a simple Lear 55, check out those altitudes too.)

The briefer will provide winds aloft forecasts for as many stations along the proposed route of flight as are necessary, and at each station, those altitudes above and below the proposed initial cruising altitude; if the initial cruising altitude is 7,000 feet, he will probably provide winds for 6,000 and 9,000 feet.

The briefer is supposed to interpolate between altitudes in order to provide winds aloft for the specific cruising altitude, but in most cases this will only be done on request. In any case, you might want to know what the winds aloft are for two or even three reported altitudes—within the range of practical cruising altitudes available—to flight plan for the most fuel-efficient cruising altitude.

A good understanding of the expected winds aloft is essential to accurate flight planning, VFR or IFR. The winds aloft determine wind correction angle (for accurate dead reckoning and VOR or LORAN tracking) and tailwind component, from which groundspeed, time en route, and fuel required are calculated.

The Weather

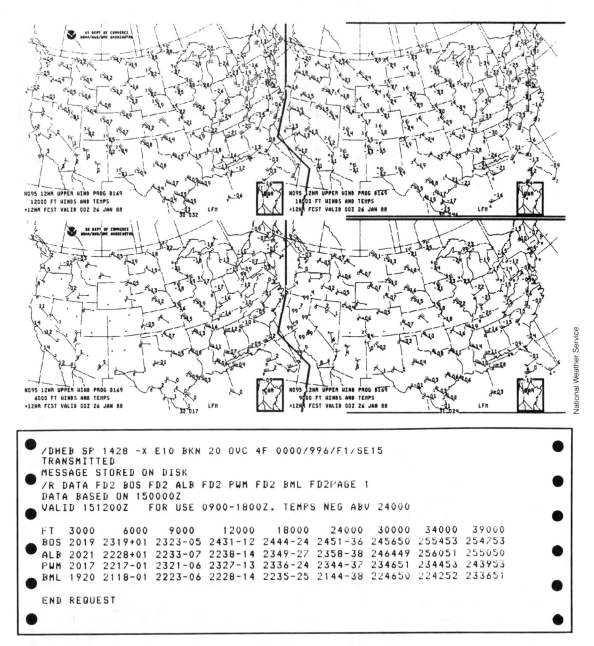

Fig. 7-6. *Winds and temperatures aloft forecasts are provided in graphic form (top) and tabular form (bottom) for a variety of altitudes and locations and for three different time periods. The graphics shown here are 12-hour forecasts for 6,000, 9,000, 12,000, and 18,000 feet. The table is a 12-hour forecast for levels ranging from 3,000 to 39,000 feet MSL. Temperatures aloft are provided for all flight levels higher than 3,000 feet AGL, which is assumed to be the same as surface temperature.*

FAA Standard Weather Briefing

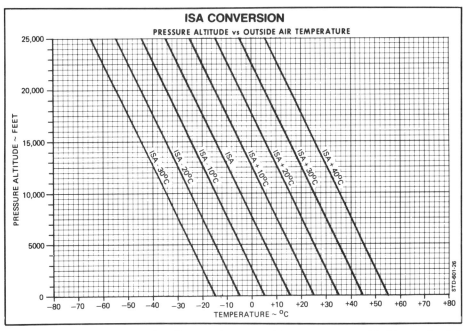

Fig. 7-7. *Temperature directly affects performance. Temperatures provided in winds aloft forecasts can be compared to temperatures provided in charts such as this to determine variance from standard. For Educational Purposes Only. Not to be used under any circumstances in the operation or maintenance of an actual airplane.*

Temperature directly affects performance. A higher than standard temperature for any given altitude translates into a higher density altitude. A lower than standard temperature translates into a lower density altitude. Because the airplane performs according to density altitude, not cruising altitude, this can be an important correction: any deviations greater than 10 degrees Celsius from standard will result in a substantial change in the performance of the aircraft.

A handy rule for calculating standard temperatures for any given altitude is to double the altitude and subtract 15, then reverse the sign; if the result is negative, make it positive, and if positive, make it negative. For instance, to figure the standard temperature for 14,000 feet, double 14 (forget the zeros), which is 28, subtract 15 and reverse the sign for −13. As you can see from FIG. 7-7, this works out to be just about as accurate as temperatures taken off a chart (the exact standard temperature [ISA] for 14,000 feet is −12.7 degrees Celsius), so this rule works quite well.

Notices to Airmen (NOTAMs)

Notices to Airmen—NOTAMs—are last minute or temporary changes to airway or airport facilities. NOTAMs do not have to be weather related and usually are not, but they

The Weather

are included as a part of the standard weather briefing because they need to be known prior to flight planning and filing. NOTAMs might dictate changes to the flight plan, and, as a practical matter, they are listed with the current weather sequence (FIG. 7-2) for the related airport. In most cases the briefer will not wait until this point in the briefing to mention NOTAMs, but will mention them with the current conditions.

In any case, NOTAMs are an important but frequently overlooked part of the weather briefing. Facility outages, unmarked obstructions, or runway closures will be covered by NOTAMs, to prevent unpleasant surprises.

ATC Delays

The last regular part of the standard briefing covers known ATC delays. Known ATC delays—in-flight delays—are fairly rare these days. Current ATC policy is to try to keep airplanes on the ground whenever there are delays en route or at the destination.

This reduces holding time and controller workload and congestion, but sometimes it does result in unnecessary delays (which the airlines don't like, so you may see this policy change back someday). If there are any known delays, though, you will be told about them, and are required, by regulation, to fuel the airplane for them (and would be a fool not to).

If you have proposed a VFR trip, this section will normally be skipped, because these are ATC delays, and ATC has no control over VFR traffic except around certain terminal areas.

Other Information On Request

That concludes the regular portion of the standard FAA weather briefing. Pilots are free to request information from flight service on military activity, long-term NOTAMs (called Class II NOTAMs), density altitude, customs and immigration, ADIZ procedures, search-and-rescue, and LORAN-C NOTAMs. These requests are covered at the end of the standard briefing.

Information on military activity is important if flying near MOAs or restricted areas. LORAN-C NOTAMs are important if the aircraft is so equipped because LORAN-C facilities sometimes suffer outages or become restricted in some way, something to know if planning on using LORAN-C for primary navigation. (OMEGA NOTAMs are also available, but very few general aviation pilots operate with OMEGA.)

The standard briefing covers a lot of ground. It is very difficult to not have either enough or the proper weather information if a pilot requests a standard briefing. With complete information it is very hard not to flight plan properly or make good weather-related decisions. It is very easy to flight plan poorly or make unsafe weather judgments when you do not have the right information.

Requesting a standard briefing or a complete briefing (the briefer will understand) is the easiest way to ensure receipt of that information.

ALTERNATES

Any deficiency in the standard briefing format is in the area of alternates—identifying those situations where an alternate is required and selecting an alternate airport that meets the regulatory requirements. Nothing in the standard briefing format specifically deals with alternates.

An alternate airport is necessary when the destination weather is reported or forecast to be lower than 2,000 and three for two hours before and after the expected time of arrival. Regardless of weather, always have an alternate, no matter what the current or forecast destination weather is. Most countries require alternates for any instrument flight, and that is a good idea.

You can never tell exactly what the weather is going to do, and it is always a good idea to have another airport as a possible destination and the fuel to reach it, no matter what the proposed destination weather is supposed to be.

The next step is selecting an alternate. The briefer has to scan forecasts looking for an airport close to the destination that, in most cases, is no worse than 600 and two (the alternate minimum requirements for any airport equipped with an unrestricted ILS). The pilot has to be sure that the proposed alternate does not have higher than standard alternate minimums (found on the back of the first Jeppesen approach plate for the airport in question).

[This takes a long time and is somewhat error prone. Perhaps information on alternate minimums could be stored in a computer, then the computer could merely scan the forecasts to see which airports were available as alternates. A list of selected alternates would help the pilot choose the most suitable alternate and flight plan accordingly.]

Identify and fuel up for a suitable alternate—preferably an airport with an unrestricted ILS and minimums to 200 and $1/2$—and make sure that it is forecast to be at least 600 and two (or higher, as specified on the back of the Jeppesen approach plate). You will always be legal this way and will always have another place to land other than the destination. Do this and you will always feel confident that the alternate will be above minimums because you have selected as alternates only those airports with forecasts better than 600 and two with full, unrestricted ILS facilities, well above the normal landing minimums of 200 and $1/2$.

TELEPHONE BRIEFINGS

You do not have to appear in person to get a good weather briefing, but it sure helps. It is, however, becoming increasingly difficult to get an in-person briefing. The long overdue modernization of the flight service stations has, unfortunately, resulted in consolidation of many smaller flight service stations into fewer stations.

Instead of most pilots having easy access to a FSS—no more than 30 or 40 miles away and often closer in "the good old days"—most pilots now find themselves farther away from an FSS than the distance to their destination, in many cases. This means that the in-person briefing has regrettably become virtually extinct.

A good briefing over the phone is possible, but it takes time and effort. Make the briefer go slowly so you can write it all down, write neatly so you can read it later, and supplement the briefing with general weather information from another source like The Weather Channel available on most cable systems. It is very difficult for a briefer to accurately and completely describe charts on the wall—the overviews, the synopses, the upper level weather charts—but it can be done, which is good, because most briefings will be over the phone from now on.

COMPUTER BRIEFINGS

One way around this is a personal computer, a modem, a subscription to one of the commercial providers of aviation weather, and a printer to provide a hardcopy. These commercial sources provide complete weather briefings, certainly more complete than what you can write down from a phone briefing, and in a form that is easier to use in the airplane. (The computer can do most of the flight planning, which usually results in an additional charge, but is well worth it in most cases, particularly when the difference means having a flight log and not having one.)

CONCLUSION

Unscientific observation has revealed that pilots get in trouble with the weather not so much because they look at a given situation and then make a bad decision, although that does happen, but because they do not get the necessary information to make an intelligent and very often fairly obvious decision.

Obtaining enough information means a thorough weather briefing and the best way to get one is to ask: "I am planning an IFR flight in a Beech F33, Boston to Richmond, Virginia, preferred routing, low level, two hours from now. I would like a standard briefing, please. I'm ready to write."

That is all it takes to get a good weather briefing. If you forget to give the briefer all the background information, he will prompt you for those parts. Forget to ask for a standard (or complete) briefing and you probably will not get it.

A good briefing is all the information needed to make intelligent decisions about the weather and flight plan accurately and thoroughly. In so doing you are very unlikely to get into situations with the weather that you cannot handle, although one way of safely dealing with that weather might be to cancel or postpone the flight.

Experience is great, especially weather experience. But it is hard to come by. It takes time to get experience and it takes experience to get experience. There is no easy way to obtain experience other than to keep trying. Regardless of your experience level, the starting point for every flight is the weather briefing. If there is an easy way to obtain experience, requesting a standard weather briefing before every flight is a good part of it.

8
Communications

Most things that a pilot does in an airplane are private; nobody else truly knows what is going on. Even in transport category airplanes, pilots only share decision-making and flying abilities with one or two other people, a copilot and sometimes a flight engineer. But when a pilot picks up a mike to talk on the radio, everybody on the frequency gets a chance to observe the performance.

Student pilots know this instinctively. That's why they have such terrible "mike-fright." They memorize perfectly what their instructor has told them to say: "Moosechip Ground, this is Waco 99 Echo, Shade Tree flight line, taxi for takeoff." They pick up the mike, position it 19 different ways in their sweaty palms, and then they say, "Ah, Ah, flight control for takeoff. Permission." They don't really know who is listening, but they assume that whoever it is knows how to do it, and they don't.

A pilot can do many things wrong when flying by himself and most of the time no one will notice. No one is there to say do this or try that or quit doing this or always do that. When solo you can usually keep your mistakes to yourself, but everybody hears you talk on the radio.

The problem is that poor communication skills is like having bad breath—nobody will tell you. You never see an experienced pilot go over to a less experienced pilot and say, "Hey buddy, I heard you talkin' on the radio back there, and I hate to tell you this, but you have a problem." However, if you flew with a seasoned captain and were responsible for the radio and could not even do that right, you would hear about it. It's the captain's flight, his "airplane," his "ticket" on the line, and you work for him, so do it his way or find yourself sweeping hangars again. It almost happened to me. My first copilot position was

COMMUNICATIONS

on a Cessna 421. I had a couple thousand hours instructing and flying single-pilot charter, but this was my first job flying with somebody else. The captain on my very first flight was a senior, gray-haired type, a former military and airline pilot. At the time he was the director of training for the company that had just hired me, and had a reputation as a nitpicker. I had my work cut out for me, but I wasn't really worried because I had talked on the radio many times with no problems, and I didn't see how this could be any different.

I received the clearance and initial taxi instructions without incident. Then the tower told us to hold short of the runway for landing traffic and I said "Roger." That was a big mistake. Captain Grayhair wheeled around: "You don't say 'Roger' to something like that. The guy in the tower has a plane on short final and he doesn't have any idea what you're going to do now, he doesn't even know if you heard him right. What if *you* think he said 'Expedite crossing the runway for landing traffic' and you said 'Roger' and are now about to cross directly in front of his traffic and even if you did hear it correctly he still doesn't know what you're going to do because all you said was 'Roger.' In fact, he doesn't even know for sure who said 'Roger,' does he? You didn't identify yourself."

Well, please pardon me. The guy said "hold short" and I said "Roger." What could be clearer than that?

Other incidents occurred, all of which memory has graciously allowed me to forget in the intervening years; needless to say, the flight did not go well. He went to the chief pilot and said I needed a lot of work and that he was getting a little tired of having to teach copilots who were supposed to know what they were doing how to talk on the radio. It was just a wonderful way to start a new job.

Anyway, as I thought about it, I was gradually able to admit to myself that he was right. I had, over the years, invented my own radio language, and nobody ever pointed out the deficiencies. I think, in retrospect, that this particular captain probably came loaded for bear when only a .22 was needed, but nonetheless the lesson has never escaped me.

Using radio to communicate absolutely essential and critical instructions is not a great idea to start with, but the alternatives, at least in any practical form, do not exist. It is therefore absolutely imperative to learn to use the system in a way that minimizes, as much as possible, the opportunities for confusion and error.

POINT OF VIEW

The most important thing learned from this mistake and subsequent "correction" was to think about the situation from the controller's point of view. He (or she) has to make many potentially unwieldy airplanes do certain things in order to keep them from running into each other; the only form of control is to issue instructions over a radio to the people around them.

The job is tough enough to begin with. The airplanes are in constant motion, they operate at different speeds and altitudes, and the instructions they can respond to are basic and primitive: go faster, go slower, go up, go down, go here, go there. As simple as these instructions are, there is still a lag in the response time and a limit to how much faster or slower they can go.

What is worse, having issued these instructions, there is no guarantee that the person driving the airplane will do what he is told. A controller, with one plane on the runway and one on short final, for instance, can issue an instruction to the airplane on the runway to "expedite clearing the runway" ("go faster"), but there is nothing the controller can do if the pilot does not expedite, except to tell the other airplane to go around.

LIMITATIONS OF RADIO COMMUNICATIONS

ATC is difficult at various times because the radio will quit working, make noises, garble words, fade, receive but not transmit, and transmit but not receive. Even when the radio does work properly, there might be as many as 20 aircraft on the same frequency, and only one person can talk at a time. One transmission completely ties up the frequency, preventing the controller from issuing his instructions, and there is nothing he can do except sit there and go crazy.

Probably the single most frustrating characteristic of using a radio for communication, from the controller's point of view, is that there is almost no way to know who is talking unless that person identifies himself.

Considering all these potential problems, aren't you glad you're a pilot and not a controller? Try to always remember what the controller is trying to accomplish—think from his point of view—and remember the limitations inherent in trying to do that with a radio; proper radio technique is so important.

DATA LINKS

Someday airplanes will have data links and the radio will be merely a backup. When the controller has instructions he will type them into a keyboard and they will be transmitted to the aircraft and displayed on a screen. A bell or light will go off directing attention to the message, and you will push a button acknowledging receipt of that particular message, which will go back to the controller's panel. If there is any discrepancy between what he sent and what you acknowledge, the controller will be alerted. The communication will be positive, clear, and verified both ways. The message will be transmitted without modulation (voice communication must be modulated, which takes a lot of power), so it will have greater range and penetrate interference better.

Comments

Data link will be a better system, but I will miss the human contact between controller and pilot, and I will miss not hearing what is transmitted to the other aircraft. The more you fly the better you get at visualizing the "Big Picture" based on all ATC communications and once in a great while you even catch a possible mistake. These problems will have to be "addressed," as the politicians say, and I'm sure they will be. It will be a better system. I wish I could get more excited about it.

COMMUNICATIONS

COMMUNICATION PROBLEMS

Problems with radio communications usually fall into one of two large categories—either a lack of clarity or a failure to verify the transmission. No matter how good radios get, a lack of clarity—communication in the literal sense—will be a problem from time to time. Even in normal speech people sometimes do not hear nor understand what is being said properly, and radios only make the problem worse. Because communication is so crucial to aircraft control, and because radios are less than perfect communication devices, there has to be some way to verify that what you actually heard and what you were supposed to hear are the same thing.

If the controller says, "Hold short for landing traffic," but you think he said, "Expedite crossing for landing traffic," that is a problem of clarity. If you respond with "Roger" to what you think is an instruction to cross the runway, the controller assumes you are responding to his actual instruction to hold short, and will not know anything is wrong until he sees you crossing in front of his traffic.

That is a problem in verification. This chapter will emphasize in specific ways, these two main points of radio communication: *clarity,* to avoid confusion and repetition, and *verification,* to catch the inevitable misunderstandings that happen anyway.

CLARITY

Captain Grayhair used to talk so slowly, and enunciate so carefully, that I was embarrassed for him. (We eventually got along fine, by the way, and I still see him from time to time.) He talked on the radio as if he were talking to a foreigner. I sure wasn't going to say anything to him, but I thought he had a lot of nerve criticizing me for the way I talked on the radio when he sounded like a "45" being played at "$33^{1}/_{3}$."

A couple of weeks later, I was flying in another airplane and heard him on the frequency and he sounded great. His transmissions were clear and distinct and, while he did talk slower than almost everybody else on the frequency, it didn't seem too slow at all. In fact it was a pleasure being able to understand someone so well.

I figured somebody with more nerve than me must have talked to him. But the next time I flew with him he was talking slowly again. And the next time I heard him on the radio he sounded good again.

You get the picture, right? (Probably quicker than I did.) To be understood over the radio, speak extra clearly and probably much slower than normal. It might sound strange as a "transmitter," but not as a "receiver." If you don't think so, make it a point to really listen the next time you fly. Fast talkers are hard to understand, especially if they also do not speak extra clearly. (And of course, the fast talk is harder to enunciate clearly.) The guys who really come across well are the guys who are actually speaking slower than normal—it ends up sounding normal over the air. (And some of those "guys" are gals.)

Probably the most common mistake private pilots make with the radio is speaking too loud. This causes their voice to rise in pitch and the result is a high-pitched and distorted sound that is hard to understand, annoying, and unmistakably stamps the pilot as an amateur. Perhaps there is a reason. Student pilots are always a little afraid of the radio and instructors are always telling them to speak up: "Hold the mike close to your mouth, and speak up!" Eventually it becomes a bad habit. Also, the transceivers in trainers are usually bottom-of-the-line, and shouting, in a vain attempt to compensate, is the inevitable result. Holding the mike close to the mouth is correct—mainly to take advantage of the noise-cancelling properties of the microphone, not to increase the volume—but speaking up does not mean shouting.

A transmitter needs vocal energy to create and modulated energy, so to a certain extent you *can* increase the power of a radio by increasing voice volume. But only up to a point and after that point distortion occurs. If you must shout to be heard, have the transceiver checked by a technician.

SIDETONES

There is only one way to learn to use the radio properly and that is to hear what you sound like, then you will know exactly what is wrong and how to correct it. One way to hear yourself speak is with a sidetone. A sidetone is an "echo" of the actual transmission that is routed through the speaker or earphones so you can hear yourself.

A sidetone is essential if you use earphones because we need to hear our own voices to speak properly; it is very hard to speak if you cannot hear yourself. (This is one reason why people who lose their hearing can be hard to understand; they still remember how to talk, but they do not get feedback on what they are saying.) Because the earphones cover ears, it is very hard to hear yourself speaking with them on, so a sidetone is provided. A sidetone is not essential when using a speaker but it is still very desirable in order to monitor the actual sound of a transmission.

A sidetone takes a little getting used to, but once you learn to listen to what you sound like without the distraction (which comes very easily, actually), you will have no problem making transmissions that are clear, readable, and properly pitched. The whining, semi-hysterical sound of the amateur will go away. This is why "airline captains" all seem to affect that low-pitched command-presence growl over the radio.

You can carry it too far, of course, but those who fly a lot have learned, mainly by listening to themselves, what it takes to make their voice as clear and readable as possible. Talking like an airline pilot might sound like an affectation at first, but no, it is the correct way to talk on the radio.

If you do not have a headset (an earphone/boom mike combination, which presumably has a built-in sidetone), talk to a radio shop about getting a sidetone wired into the speaker system. A new audio panel might be required, which can be expensive, so the headset route might be the better way to go (FIG. 8-1). A headset with a push-to-talk switch is probably a better idea in a single-pilot operation anyway, because it is easier to hear the transmissions with earphones, and you don't have to reach for a mike.

COMMUNICATIONS

Fig. 8-1. *Audio panels, such as this King KMA 20, provide for convenient selection of transmission and receiver sources. Most also provide for a transmission sidetone through both the headphones and the speaker, a great aid in monitoring transmission quality.*

THE LANGUAGE

Careful transmissions should take care of the physical aspects of clarity: speak slowly and listen to adjust your tone and pitch as necessary. Another aspect of clarity is using the language in a consistent predictable manner. No matter what you might tell your friends and companions that say, "I just don't know how you understand a *thing* they are saying on that little, bitty radio," the fact is it *is* occasionally hard to understand. Predictable and consistent language is easier to understand.

It is possible to decode Morse code as fast as 10 words per minute merely by memorizing the dots and dashes that go with each letter. A person cannot go faster than that without breaking "The Barrier," which is the translation process. You hear a pattern of dots and dashes, recall just as fast as possible which letter goes with that pattern, write it down, and hope that was fast enough not to miss hearing the next letter. The only way to decode faster than 10 words per minute is to learn to recognize the letter merely from the sound of the dots and dashes without stopping to remember what it is. Presumably a person can quickly move up to 20 words per minute or more.

The same thing happens in speech. When someone says something unexpected, you have to translate—pause for a second and reconfigure your thinking to what was actually said, instead of the pattern you expected—and this slows things down and leads to misunderstanding and error. Writers often rearrange words in unexpected patterns on purpose to reveal new meanings. Talking to ATC is not poetry, it is the opposite of poetry: communication in the most basic and literal sense—do not play games with words here.

Radio talk actually is a language of its own. It is very similar to English (thank goodness for small favors) and has a very simple grammar and vocabulary, but it is still a specialized language that must be learned like any other language. When controllers are training, this is probably the most important part of the training, and using nonstandard terminology is a serious error.

A controller trainee might fumble with the language because he is not yet fluent. It is disconcerting to hear a controller say something "nonstandard" and it makes you wonder if that is the only thing he has messed up, then you doubt whether you genuinely understood what he meant. On the other hand, the controller who used standard, predictable terminology has an "air" of control and assurance; assume that controllers react the same way to pilots.

The best way to learn standard terminology is listen to pilots who do it correctly. Avoid slang and shortcuts. Do not get cute. You might not appreciate saying "affirmative" or "negative," and people who talk like that in everyday speech usually are a little offensive, but that is correct in aviation. Not only is it clear and unmistakable—"affirmative" does not sound like anything else—it is also predictable. It will not cause the controller to mentally stop for a microsecond and translate what you really said into what he expected.

To learn the language of aviation radio communications, listen, imitate, and practice. Listen to pilots who seem to know what they are doing and act like they are in charge of the situation, therefore earning respect and cooperation from controllers.

This is one area where you do not have to actually fly with the "old guys" to learn. Listen to other transmissions on the radio and learn from example. Also listen to the controllers; they have to use standard terminology, so they are the models for "radio speak." (Of course controllers sometimes deviate from standard, and Greg Norman sometimes goes over the top on his backswing; but you have to know the right way before you can get away with doing it wrong.)

KEY WORDS

Here is a short list of the most useful words, with some comments on how to use them.

Roger: (Might as well as take the hardest one first.) Roger does have its place, but

COMMUNICATIONS

only in nonessential communications. It basically means "okay." (The Pilot/Controller Glossary in the *Airman's Information Manual,* says it simply means, "I have received your transmission.") If a ground controller says the aircraft in front of you will move up, "Roger" would be an acceptable reply. Or if he said, "You can take it to the end of the runway," you could say "Roger." But "Roger" should never be used in place of "affirmative" or "wilco" in essential communications, nor should "Roger" be a shortcut for not repeating a clearance. The proper reply to "99 Echo, taxi up and hold short" is "Taxi up and hold short. 99 Echo." Using "Roger" is like calling the boss by his first name; if you don't know for sure that it is okay, do not do it.

Standby: This very useful word provides the pilot with some of the control over the communications. If you are busy, or don't know the answer at that moment, say "standby." If the ATC request is critical and cannot wait, he will let you know. Nine times out of 10 it will not be critical and a later, well thought out, accurate answer is much better than a quick, hasty one.

Occasionally a controller or flight service station specialist will ask something inconsequential right after takeoff. You are busy flying the airplane, and do not want to be distracted at that moment; if you don't say anything, he is going to bug you again. A quick "standby" usually gets the message across: "Don't talk to me when I am busy trying to get this thing into the air, unless it is awfully important."

Once in a great while someone will even call you with something nonessential while *rolling* on takeoff; this is inexcusable because the first reaction should be to abort as soon as you hear the call sign on the takeoff roll. Normally no one calls you on the takeoff roll unless something is very wrong, and the sooner you start stopping the better. This happened to an Eastern 727 at Washington National. Eastern was in position and had been cleared to go, but was slow in rolling. Just after the 727 did start to roll, the controller said "Eastern 123 (not the real numbers), you *are* cleared for takeoff you know." The 727 was aborting as the controller spoke. The captain heard "Eastern 123" and that is apparently all he needed to hear to abort. The tower said again, "I repeat "Eastern 123 is cleared for takeoff." All the captain said was, "Too late now."

This was more of a screw-up than anything else. The tower controller probably could not tell that the 727 had started to roll. But it illustrates the point: If a controller calls you on the takeoff roll, assume it is a critical communication and act accordingly, not "standby." Whatever you do at this point, talking on the radio is not part of it. If it turns out that the call was *not* essential, a telephone call to the tower might be in order. We are all human and controllers make mistakes, too. A reasonable explanation of the problem encountered when someone calls on the takeoff roll would not be out of line.

Unable: This little word can cover a lot of ground and shorten a transmission. "We're not going to be able to use the short runway;" *"Unable the short runway."* "I couldn't raise anybody on 132.5 . . . you got another frequency?" *"Unable 132.5."* "Cherokee 8 Fox Lima is a little too heavy for Flight Level 350 today;" *"Unable Flight Level 350."*

Over: To be used rarely. The only purpose of the word "over" is to signal the end of a transmission and the end is usually very obvious and does not need this embellishment.

Verification

The aircraft call sign works much better and also provides positive identification. For instance, if the controller says, "99 Echo, climb and maintain niner thousand," call back "Out of eight for nine, 99 Echo." Notice that the controller did not end his transmission with "over," and if he had, it would have sounded very strange.

About the only times you might need to use "over" would be after a long pilot report or at the end of a flight plan or maybe a long request to make sure the person at the other end knew you were through. You also might want to end the transmissions with "over" when using high frequency radio for position reports. (High-frequency—"shortwave radio"—is often used on long, overwater routes.) High frequency can be very hard to understand, and "Over" is sometimes necessary to make sure the person you are talking to knows you are done. But these times will be rare. If you are using "over" a lot, you have probably picked up a bad habit.

VERIFICATION

So far everything has dealt with clarity: speak slowly and clearly, do not shout, use standard terminology. The other half of the picture is making sure that everybody understands everybody, because aviation radio transmissions are too important to leave to assumption. ("Assumption is the mother of all mistakes." Actually, we did not say "mistakes" in the army.)

All clearances should be verified. (A clearance is not merely the IFR route clearance copied while on the ramp prior to takeoff. That is the initial clearance. Any instruction from ATC is, technically, an amendment to the initial clearance, and is therefore also a clearance itself.) Verification means reading back the clearance. The regulations do not specifically require a readback, but you *are* required to "request clarification" if the meaning of a clearance is uncertain, and under IFR you must "receive" an appropriate clearance.

It is all right to abbreviate the readback, as long as the essential words are included. If in doubt, it is always better to err on the side of reading back too much, and there is nothing wrong with reading it back word-for-word. Thus, if the ground control says "99 Echo, taxi to runway 19," you might read back "Taxi to 19. 99 Echo." But if ATC says "99 Echo, call holding short of Golf taxiway," you should reply, "Call holding short of Golf. 99 Echo."

There is only one word in that clearance that is not essential and does not need to be verified: taxiway. The controller must have a reason for not wanting you to go any farther than Golf, or he would have cleared you to the runway. He also has some reason for wanting to know when you get there, and he wants to make sure 99 Echo and not somebody else gets the message. So the only proper readback in this case is, "Call holding short of Golf. 99 Echo."

An accepted procedure is calling the "tower" after conducting the pre-takeoff checklist. You will most likely get one of the following clearances: "Taxi up to but hold short of the runway," "Taxi into position and hold," "Cleared for takeoff." The standard answers are, respectively: "99 Echo, up to and hold short," "99 Echo, position and hold," and

COMMUNICATIONS

"99 Echo, cleared to go." (The last one is not technically proper of course—you really should reply "Cleared for takeoff"—but it is completely acceptable.) The aircraft call sign can go at the beginning or the end of each of these transmissions: "99 Echo, cleared to go," or "Cleared to go, 99 Echo."

At some point on climb-out the controller will tell you to contact departure control. He usually will not provide the departure frequency because that should be part of the clearance or SID (standard instrument departure).

When ATC says, "99 Echo, contact departure," reply "99 Echo. Good day" and switch frequencies.

En route, most clearances will be for altitudes or headings. The custom on altitudes, going back to the days when control was based entirely on aircraft reports, is to report leaving the old altitude for the new altitude, but to omit the report reaching it unless requested by ATC.

If the controller clears "99 Echo, climb and maintain one two thousand," you would read back, "99 Echo, out of one one thousand for one two thousand." When the controller clears "99 Echo, climb and maintain four thousand. Report reaching," you say "99 Echo, out of three for four. Call reaching." When you write a "call reaching" clearance down, put a circle around the altitude as a reminder to call ATC when level.

Any time you get a heading change, read it back because it is just as important as an altitude change. If ATC clears "99 Echo, left heading 270," read back, "Left 270, 99 Echo." If ATC clears "99 Echo, turn 30 degrees to the left, vectors for traffic" say, "99 Echo, 30 degrees left." Comply with the request unless safety is compromised by unidentified traffic, terrain, an obstruction, or weather; tell the controller what the problem is and request another clearance.

After awhile, when you get used to reading back all the essential numbers and words, it almost becomes a litany: ATC says this, you say that. This is not to say you merely blindly repeat everything he says; obviously you have to know what he means and you have to comply with it. But there is a sense of assurance and confidence in the system that comes with knowing how to use the language properly. Listen to the "airline voices" and try to sound like them. Listen to corporate and charter pilots, too. You will hear good and bad, depending on the company policy and the flight department "personality." Try to see why some pilots seem to communicate easily, and others, while being possibly entertaining and original, seem to interfere with the basic job at hand.

Above all, avoid slang and do not get cute. You might be a really wild and crazy person, and that is just the way you are and the way you like to be, but when all is said and done, this is not playtime, this is work. This is the job of getting from one place to another using a form of transportation known as an airplane, and there is a right way to do that job and a wrong way.

Before breaking any rules, make sure you know them. When you have a couple thousand hours of instrument cross-country time, maybe then you can take the liberty to chit-chat and break a rule once in awhile, but the more you fly the less fooling around you will probably do.

FREQUENCY MANAGEMENT

Presumably every airplane has two transceivers for communication, even though it is legal to operate with one. Radio communication is too important, and radios fail too often, to even *think* about operating IFR with only one radio. (The L-1011's I fly have three, but that is mainly because "mother" always wants to know where we are; one radio is always set to company frequency; we cannot have any fun.)

The ideal arrangement is to have at least one radio with a frequency preselect feature. On the ground, you can "lead" your frequencies by one because you know what the next frequency will be. (They are on the approach plate.) If you are talking to ground, the next frequency will be tower; by setting that frequency in the preselect window, all you have to do is switch frequencies. Likewise, once on tower, set the departure frequency in the preselect window.

Frequencies will be assigned after departure and cannot be anticipated. Use the preselect window for storage of the *old* frequency. When the controller says, "99 Echo, contact Kansas City Center on 132.7," you say "Thirty-two seven. 99 Echo," Merely set 132.7 in the preselect frequency, "swap" the frequencies and report in on 132.7. If you do not get a response on 132.7 (try three times with at least 30 seconds between) "swap" the frequencies again and report to ATC "99 Echo, unable 132.7." This saves having to write down and cross out used frequencies.

If you ever need to go back to your old frequency and forgot to write it down or improperly stored it, do not panic. Two things can be done: call center or an FSS. Look on the en route chart for the appropriate center frequency in that area, or at least an active frequency nearby. As long as you can get a hold of someone in that center's area he or she can direct you to the right frequency or call flight service and explain the situation. They have direct landlines to center and can give center a call and straighten it out. It happens to everybody occasionally, but if it happens a lot—more than once a year—you are doing something wrong.

What do you do with the second radio aside from having it as a backup? This is an area of personal preference, but common practice is to use the second radio for non-ATC functions: flight service, unicom, a discrete FBO frequency. In remote parts of the world, it is common to tune the number two com to 121.5, and overwater it is mandatory. Dial in 121.5 occasionally, perhaps to help someone or hear interesting transmissions. It is a good idea to segregate the uses: number one radio for ATC, number two for everything else. That way you always know the radio assignment.

This is a much better method than trying to switch back and forth from the number one radio to number two, even without a preselect feature. If you try to alternate radios, it is very easy to get confused about which frequency is active. The number two radio is a backup and any other miscellaneous use is fine, but do not try to integrate it with normal ATC communications.

COMMUNICATIONS

SQUELCH AND VOLUME

The volume of certain transceivers cannot be turned down all the way; they can be turned off, but no matter how low the volume is turned you can still hear *something* that would eventually get your attention. This is a terrific feature because the volume does sometimes get accidentally turned all the way down and ATC can go crazy trying to reach you before realizing it has been a long time since hearing anything. Check the squelch (FIG. 8-2). If the volume was turned down too low to hear, or the radio has quit working, you will not hear anything with the squelch turned off or overridden, if automatic. If the volume *was* turned all the way down, immediately reestablish contact with center: it is not the end of the world, but an entire center might be doing handstands trying to work around you and the sooner things get back to normal, the better.

If the transceiver volume is okay, wait a moment or two longer—usually another pilot will ask why it is so quiet, saving you the trouble. If you cannot stand it anymore, request a radio check. Nine times out of ten the controller will come back "Five by five, how me? If ATC does not answer you, ask if anyone else on the frequency can relay a message; if someone answers, explain the problem and see if he can get a better frequency from ATC. If *nobody* answers, try another radio. If that doesn't work, try FSS on 122.2, first on that radio, then on the other, and then on 121.5.

If that does not work, turn up the volume on the VOR navigation receiver because an FSS might be trying to relay messages that way. If nothing works, and you are in visual meteorological conditions (VMC), land and cancel. If in IMC, continue as previously cleared, according to communications failure procedures described in the FARs and the AIM. If you haven't thought about those procedures since the written test, review them; *now*.

MEMORY TRICKS

When ATC requests two things at once, I can handle both without too much trouble. For instance, "99 Echo, turn 30 degrees to the right. Climb and maintain one zero thousand." I can usually remember both of those things long enough to say, "Thirty degrees to the right. Out of nine for one zero thousand. 99 Echo." Once I say it out loud, it sticks pretty well in my head, although I immediately reset the heading bug 30 degrees and the altitude alert to 10,000, or write it down.

Three things—"Turn 30 degrees to the right. Climb and maintain one zero thousand. Contact Memphis Center on 135.5"—it gets a lot tougher. [One memory trick is to read it back in reverse order. You are supposed to read it back in the same order, but I read somewhere, and it seems to work, that the mind can remember a list of things better in reverse order than in the original order. The controllers prefer a read back in the same order, but it is not required, and I have never heard a controller complain.

If you don't like changing the order, read back what you can and ask for the rest: "99 Echo, right 30 degrees, out of nine for one zero thousand, say again Memphis Center." If

Memory Tricks

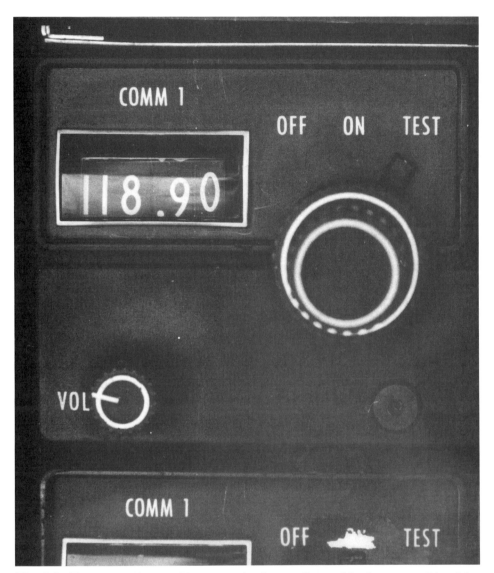

Fig. 8-2. *The test position on this communications radio provides for an override of the automatic squelch feature, a useful way to verify not only proper operation, but also volume setting.*

there are any controllers reading this, I personally think you'll save time in the long run by not giving out more than two instructions at once. It goes a lot smoother.]

Another memory trick is to visualize the numbers in midair as the controller says them. It might not work for you, but I have found that I can hang onto numbers better

COMMUNICATIONS

when visualized this way, rather than memorize them long enough to reply or find a pencil and write it down.

SIMULTANEOUS TRANSMISSIONS

Listen before transmitting. Airborne VHF is FM—frequency modulation. There are many advantages using FM instead of AM (amplitude modulation), but one *disadvantage* is that two people cannot talk at once. When they do, the two transmissions cancel out, resulting in a very loud squeal and neither transmission will be received by the intended recipient. It creates all kinds of confusion because the parties involved do not know there is a problem. This usually leads to more confusion as everybody else tries to help them out—an Alphonse and Gaston act.

If two aircraft call at once, let ATC straighten it out, do not add to the confusion with your own two bits. But if you know ATC was blocked, sometimes a quick "blocked" will do the trick. This is short enough to not wipe out any reattempts, and lets ATC know right away that he did not get through. The important point is to *listen* first and make sure no one else is talking before you do, which helps prevent simultaneous transmissions.

Once in a great while two people block each other by chance; two people listen, hear nothing, and both decide to start talking at exactly the same moment. It would be nice if the transceiver manufacturers could invent a listening device that would alert a pilot to this condition. It is a difficult problem because it is hard for an airborne transceiver to receive and transmit at the same time—the transmission would destroy the receiver.

Most of the time simultaneous transmissions are avoidable. Most of the time they happen because somebody just switched frequencies and started immediately talking without listening or the volume was turned down. Sometimes it happens because someone down low keeps trying to establish radio contact before flying in range of ATC. He cannot hear ATC talking so he keeps calling, wiping out everybody else within range. Determine the reception range and wait until you get up there to start calling. A lot of this is courtesy and common sense. Remember, this is a big party line, and not a very good one at that, and it takes a lot of cooperation on everyone's part to make it work.

RADIO PROCEDURES: THE KEY TO THE SYSTEM

The point of proper radio procedures is clarity and understanding—communication that is positive and absolute. The backup to prevent serious errors resulting from inevitable misunderstandings is verification. The goal is fluency with the radio vocabulary. The way to get that fluency is to listen, imitate, and practice. Fluency with the radio leads automatically to fluency with the ATC system as a whole because the communication system is its heart and soul. The pilot who knows how to communicate is well on the way to mastery of his craft.

9
Emergency and Abnormal Procedures

A PILOT'S JOB HAS TWO PARTS. ONE PART IS FLYING THE AIRPLANE. THE OTHER IS BEING prepared to deal with emergency and abnormal situations. Since major malfunctions are fairly rare, dealing with abnormal and emergency situations is usually seen as separate from the normal task of getting the airplane safely from A to B, but the two tasks are not separate. Being able to deal with emergency and abnormal situations at any given moment is an integral and extremely important part of the pilot's job, but, for better or for worse, as long as airplanes continue to become more and more reliable and failures continue to be less and less common, this apparent separation is inevitable.

The bulk of this book is concerned with flying airplanes safely and routinely from A to B. But "flying airplanes safely" also means being prepared for the nonroutine, and that is what this chapter is all about.

When a system malfunction occurs, the pilot must analyze the situation, reconfigure the aircraft to minimize further damage or danger, and optimize the remaining performance and capabilities of his aircraft. In the early days of aviation, when aircraft systems were relatively simple, the pilot was expected to know his aircraft well enough to be able to handle any failures or problems by relying on a combination of memory, experience, and ingenuity.

But as airplanes became increasingly complex, total reliance on the memory, experience, and ingenuity of the pilot became unacceptable. The systems became too complex to reasonably expect the average pilot to be totally familiar with each and every detail, and the adverse consequences of faulty procedures became unacceptable.

EMERGENCY AND ABNORMAL PROCEDURES

Failure, for instance, to bring a functioning generator back "on line" in an open cockpit, VFR-only aircraft, was of little consequence, but as aircraft became increasingly dependent upon a reliable source of electrical power for navigation, communication, lighting, and the control of other systems, a consistent and reliable solution to the temporary loss of that power became imperative. In an "all-weather" environment there are no inconsequential systems.

Not only is complexity a problem, but people can be a problem too. Not all pilots are equally experienced or ingenious, and relying on these qualities to solve whatever problems arise often leads to more problems. There will always be a place for a human being as the final arbiter in an aircraft, but most problems can be anticipated, and most problems have an optimum solution that is best determined ahead of time.

CHECKLIST ORIENTED PROBLEM SOLVING

The aviation community has attempted to deal with these combined problems of complexity and human variability since the white silk scarf left the cockpit. Lead by the military, the airlines, and the better corporate flight departments, a two-part strategy based on detailed emergency and abnormal procedures checklists, and backed up by regular and thorough training in the use of those checklists under simulated failure conditions, has evolved.

The pilot of a complex, transport category aircraft is no longer expected to be able to resolve problems as they arise based on his ingenuity alone, nor is he expected to be able to react solely from memory to each and every possible failure. He is only required to solve those rare but theoretically possible problems that cannot be anticipated. He is only expected to be able to react from memory to those items that are so critical that his response must be virtually instantaneous; items that cannot even wait for a checklist to be picked up and referenced. Very few items in an airplane are that critical, but those that are must be identified, memorized, and practiced.

TWA L-1011s have 16 emergency checklists, but only three of those checklists—engine failure, engine fire, and tailpipe fire—have memory items, and they are all the same item: close the throttle. Everything else waits for the checklist. This reduces pilot workload to an acceptable level, regardless of the complexity of the aircraft, and it results in a much higher rate of successful problem resolution for all aircraft.

This doesn't mean the pilot has become a robot—no checklist can ever be made complete enough to cover every possible contingency. The pilot still has to have a general understanding of the systems in order to handle those situations that do not fit a checklist. It does mean that a large part of the troubleshooting and repairman functions of the pilot have been replaced by the use of checklists, freeing the pilot to do what he does best, which is fly the airplane.

So far this philosophy of checklist-oriented problem-solving has not filtered down to a significant extent below the airline and corporate turbine level. Most general aviation pilots have read the manufacturer's owner's handbook for their aircraft and most pilots have a general idea of the aircraft systems and the manufacturer's recommendations for dealing with various emergency and abnormal situations. Very few are aware of the need for specific checklists for each emergency and abnormal situation that can be anticipated.

This is too bad because the typical general aviation aircraft today is as complex as the airliner of the previous generation when the emphasis on the use of checklists for problem solving began. Personal aircraft have reached the point in sophistication where they require checklists for emergency and abnormal situations. The general aviation pilot that does not have access to regular simulator training and systems ground schools especially needs something to count on in a crunch. The best device invented for that purpose is a checklist.

CHECKLIST PREPARATION

Unfortunately it is impossible to provide you with a set of checklists that will cover each and every system for each and every airplane—airplanes vary too much. But I can give you an idea of what areas should be included in any set of emergency and abnormal checklists, and an idea of what items should be included in each area. With this information for background, you should be able to go to the owner's handbook and flight manual and create systems checklists based on the information contained in those publications.

If you own or fly a fairly modern aircraft—one manufactured within the last 10 years or so—the operating handbook and flight manual will follow the GAMA (General Aviation Manufacturers Association) "Handbook Specification Number 1." "Specification Number 1" is a standardized format for all aircraft information, and is a tremendous improvement over the mishmash of information that was provided prior to this specification. If an aircraft has such a manual, the job will be fairly simple because this format prescribes a separate section for emergency procedures. You might want to mark the memory items with red ink and perhaps add tabs to help locate specific problems, but the bulk of the work in setting up a series of emergency checklists will already have been done.

If you do not have a centralized location in the owner's handbook for this information, the best thing to do is to create personal emergency and abnormal situations checklists. It is a lot of work, but there is no better way to get to know aircraft systems. You'll be glad you have done it if you ever have to use one.

Physically, these checklists don't have to be anything fancy—a small, loose-leaf binder with plastic inserts for each of the emergency and abnormal situations, appropriately tabbed for quick referral, will do nicely. Or you might be able to get everything on the front and back of two pieces of heavy paper, one for emergencies, with red borders, and another for "abnormals," with black borders, and have them laminated for endurance. Whatever works. The only physical requirements are something to store within easy reach—like in a side pocket or under the seat—and be easy to use. In a true emergency situation, stress is going to be a significant factor and you want to get to the checklist without difficulty, and locate the appropriate part.

EMERGENCY AND ABNORMAL PROCEDURES

In creating checklists, it is very important to transfer the information from the appropriate parts of the flight manual and owner's handbook accurately and without "editorial license." The temptation is awfully great, as long as you are making up a checklist anyway, to make your own changes too. If you do make changes, you are wandering in the area of the unknown—the realm of the test pilot.

Follow the flight manual exactly (because those are legal limitations) and avoid the temptation to make changes to the manufacturer's recommendations. Very often there is more to a specific recommendation than meets the eye and making what seems like a perfectly simple and obvious change for the better might come back to bite you later. If you really think you have a better way, write the manufacturer and see what he says, but be very careful about making changes on your own.

EMERGENCY VERSES ABNORMAL

It is important to differentiate between emergency and abnormal situations. The difference between an emergency and an abnormal situation is that, in an emergency situation, the safety of the flight is in immediate jeopardy, and generally some form of corrective action must be taken immediately, often from memory, if not instinctively.

An abnormal situation is simply something that is not a normal occurrence. An abnormal situation requires corrective or compensating action, but the safety of the flight is not immediately jeopardized and time is not so critical.

Generally, in an emergency situation, you use the checklist after the fact, both to *check* that you have accomplished the memory items correctly and to *clean up* the non-critical items; in an abnormal situation, you have time to get the checklist out and use it as a guide from the beginning.

The operating handbooks tend to call anything an emergency, but this isn't technically correct and has several disadvantages. Emergencies are important enough to be kept separate from abnormalities, especially because in an emergency, the appropriate response or responses very often must be memorized. Abnormals deserve their own section too, mainly to keep them manageable. There can be a lot of abnormal situations and it's nice to keep them separate from the truly critical situations.

This is the way it is usually done with Transport Category aircraft—emergency checklists in one group and abnormal checklists in another—and you would be well advised to do the same. Situations in this chapter will be divided likewise.

MEMORY ITEMS

The expect a pilot, under stress, to do anything from memory is asking a lot, so it is important to restrict the memory items to only the most critical elements of real emergencies. The following are generally accepted as real emergencies: engine fire, engine failure, propeller overspeed, cabin smoke or fire, electrical smoke or fire, electric trim runaway, forced landing, ditching, rapid depressurization (pressurized aircraft), emergency descent, failure of all generators, and spins.

Emergency Checklists

Not all of these will pertain to all aircraft but this list should cover all the emergency situations for most aircraft. These are not the only things that can go wrong with an airplane, of course, but in each of these situations, either the immediate safety of the flight is jeopardized, or memory items are required.

It is possible that a particular aircraft has other systems or characteristics that could result in other emergency situations. Use the flight manual as a guide, and if you wonder whether a particular situation is an emergency or not, ask if the situation jeopardizes the immediate safety of the flight and/or requires immediate action from memory. If the answer is "Yes" in either case, then it is an emergency.

EMERGENCY CHECKLISTS

I want to talk about each of these emergencies specifically and in some detail, so that you can go to the flight manual and handbook and have a pretty good idea of what to include in emergency checklists. I have also included a hypothetical checklist for "Engine Fire" to use as a model in creating personal checklists.

Engine Fire

If this isn't an emergency I don't know what is. Not only is the engine not going to work very well if it is on fire, but fuel runs throughout the aircraft, creating an ideal situation for setting the entire airplane on fire. (One advantage of using kerosene for fuel—Jet A—is that it doesn't vaporize very well, while avgas is extremely volatile, even explosive.)

The engine failure part is the least important part of this emergency. The first priority is to put the fire out and that means shutting off the supply of fuel (there isn't much of anything else in an engine that will burn). It might also be possible to snuff the fire out with increased airspeed. Check the manual for the recommended procedure, but it will probably involve putting the mixture to cutoff and the fuel selector to off, and these fuel cutoff items should be done from memory. Then there will be cleanup items that go on the checklist after the memory items: mags and master switch off. Something electrical probably started the fire. (But remember that you might have to turn the master on momentarily to transmit a MAYDAY message.) For single-engine aircraft, the checklist should include a referral to the forced landing checklist, because that is the next problem. For twins the checklist should refer to the engine failure checklist to ensure that the failed engine is properly secured.

Figure 9-1 is a typical emergency procedure checklist for engine fire. Use it as a model only because the chances of it fitting your aircraft exactly are slim. I made it up simply as an example to show a suggested format and to illustrate the general principles of emergency checklist preparation. Develop this for your specific aircraft, for each emergency situation.

Items marked "PHASE I" are memory items. "PHASE II" are nonmemory items to be accomplished using the checklist, after having checked to be sure that all the memory items have been taken care of. (That's why they call it a *check*list.)

EMERGENCY AND ABNORMAL PROCEDURES

Phase I

1. Mixture—IDLE CUTOFF
2. Fuel Selector Valve—OFF
3. Heater/Defroster—OFF

Phase II

4. If fire continues, increase airspeed.
5. Notify ATC.
6. Magnetoes—OFF
7. Generators—OFF
8. Battery—OFF

Single-engine aircraft: Refer to forced landing checklist.
Multiengine aircraft: Refer to engine failure checklist.
Land immediately.

Fig. 9-1. *This is a typical emergency procedures checklist for engine fire. Phase I items are memory items; Phase II items are checklist items, to be accomplished after Phase I items have been checked complete.*

Engine Failure

The most important consideration in any engine failure situation, for any airplane, is to maintain control of the aircraft. I'm not going to go into basic flying techniques in this chapter because that is a job for a flight instructor. But I do want to help you develop a checklist so that after the engine failure and when the aircraft is stable and under control, you can do whatever you can to restore power. And if that is impossible, a checklist to clean up the aircraft for either unpowered or single-engine flight.

For both singles and twins (nontransport category), if there is any question whatsoever about the performance of an engine prior to liftoff, abort. I don't think you really have to have a checklist for this, but you should know what configuration the manufacturer recommends for maximum braking effectiveness.

For a single-engine airplane, if the engine fails after liftoff, the most important consideration (after maintaining control) is to attempt to restore power. If smoke is coming out of the engine, or the cowling and windows are covered with oil, or there is any other sign of a mechanical failure (like a hole in the cowl where something exited), you obviously are not going to be able to restore power. But if the engine seems to have quit for nonmechanical reasons, you want a checklist to help you try to get it going again.

Fuel and ignition are immediately suspect as culprits. Check the manual and see what the manufacturer recommends; turning on boost pumps, switching fuel tanks, and mixture to full rich are standard responses. For ignition, try the left and right magnetos separately, or whatever the manufacturer recommends. You won't always have time to refer to the checklist if this happens, so these all should be considered memory items.

Emergency Checklists

When you do have time, refer to the checklist to make sure you haven't forgotten anything. A checklist is particularly helpful after a partial power loss. With a partial power loss you probably *will* have time to dig out a checklist, and a partial power loss is also fairly likely to respond to corrective action such as switching tanks, turning boost pumps on or off, and isolating magnetos.

For a multiengine airplane, the emphasis after an engine failure is on configuring the aircraft to optimize the performance of the remaining engine. Getting the dead engine started again isn't nearly as important as maintaining control and gaining or at least maintaining altitude with the remaining engine. Again, airplanes vary as to the best way to clean up the aircraft, however, in general, the gear should come up as soon as a positive rate-of-climb has been established, the flaps should be retracted as soon as altitude and airspeed allow, and the engine should be shut down and feathered as soon as control is reestablished. Refer to the manual for the specific recommended procedure for the aircraft.

The engine failure checklist for a twin will be mostly memory items except for the final "cleanup" items, such as turning the fuel pump and generator off on the failed engine and crossfeeding procedures. It is very important to have the entire procedure on the checklist for two reasons: one, it will help fix the procedure in your memory, and two, once things are under control, it is very important to go over the checklist from the top to make sure nothing has been overlooked.

In the excitement, it is awfully easy to overlook even major items like raising the flaps, and it is nearly impossible to remember all the cleanup items. You probably won't remember to close the cowl flaps on the dead engine, for instance, and shouldn't be expected to. If "Cowl Flaps, inoperative engine—CLOSED" is on the engine failure checklist, you won't have to remember to close the cowl flaps or any of the other nonessential cleanup items. The only thing to remember is use the checklist.

The last part of the checklist should be a reminder of any accessories unique to one engine and the consequences of losing them. For instance, some airplanes have only one hydraulic pump. If the engine that drives that pump goes, the hydraulics also go. What are the consequences of losing hydraulic power? Will the gear still retract? What other systems use hydraulic power?

If you lose an engine, you also lose the generator once the prop is feathered. What are the consequences of reduced generator capacity? These things should all be on the checklist somewhere—do not count on your memory. In an emergency situation, your memory will be reduced to primordial innocence.

Propeller Overspeed

The rotational stress placed on a prop is directly proportional to the speed at which it rotates, and the stress increases exponentially—a small increase in prop speed results in a large increase in stress. If the prop control fails, the blades will normally fail to flat pitch and the prop speed will increase. If this happens at cruise, the prop RPMs will usually zoom past the redline. The prop won't stay together long at RPMs over redline (take a good look at what's holding the blades to the hub during the next preflight), so immediate

EMERGENCY AND ABNORMAL PROCEDURES

action is necessary. If the prop control is broken, you won't be able to reduce the RPMs with that, but you can reduce power and this should be done immediately and from memory. You might have to bring the power back almost to idle before the RPMs drop below redline, but that's less of a problem than having the prop come apart.

Once the RPMs are under control, you should have time to refer to the checklist, which should tell you to check the oil pressure, the lack of which probably caused the prop control to fail and the propeller to overspeed. Then the checklist should provide a note to see the checklist for forced landing for single-engine airplanes, or possible precautionary shutdown for twins.

Cabin Smoke or Fire

The greatest danger in situations involving smoke or fire is usually asphyxiation and smoke blindness. Fires can be extinguished, but smoke is much more difficult to deal with. If you have oxygen available, your first reaction to smoke of any sort should be to don the mask. Then, if you have smoke goggles get those on. Now you can breathe and see.

If you don't have oxygen or smoke goggles, the only thing you can do to clear the smoke is to increase the ventilation and hope you can see and breath well enough to find the checklist and isolate the source so the smoke will clear. If that doesn't work think about an emergency descent: get the thing on the ground before it gets any worse.

Smoke goggles are important because your eyes will involuntarily close if exposed to enough smoke, and wild horses won't get them open. Picture flying along in an airplane, the smell of smoke everywhere, and your eyes are closed and won't open. If you can't locate smoke goggles through normal aviation supply channels, try safety goggles with the ventilation holes taped over. Every airplane should have a set within reach of the pilot.

The checklist should remind you to use the fire extinguisher if flames are visible (in the excitement, it is hard to remember that you even *have* a fire extinguisher), and to increase the ventilation if there is smoke. The checklist should then ask you to determine whether the fire or smoke is electrical or nonelectrical in origin, and divide into two parts, one for each of these situations.

Electrical Smoke or Fire. If the smoke or fire is electrical in origin, attempt to isolate the faulty component. Sometimes the source will be obvious, but if it isn't, the basic method for finding it is to turn the master electrical switches off (battery and generators), and pull all the circuit breakers. Then turn the electrical switches back on. Nothing should happen with all the breakers pulled, but if the smoke returns, turn the switches off. (Occasionally there are one or two electrical items protected in some fashion other than with circuit breakers and one of those items *could* be the source of the fire.) Assuming nothing happens when the electrical switches are turned back on, reset the breakers in order of importance until you smell or see smoke again—that's the bad circuit.

The checklist should also remind you to think about terminating the flight early unless you are certain you have isolated the faulty system and can operate safely without it. (This isn't out of the question at all. Suppose a radio burns up. Once that circuit has been isolated and the smoke has cleared and ATC has been notified, there is no reason why the flight cannot be continued.)

Nonelectrical Smoke or Fire. If the smoke is nonelectrical in origin, it will most likely be environmental: heating, air conditioning, or pressurization. The checklist should remind you to try to determine the source and either put it out, turn it off, or isolate it, normally by turning off all fans and air outlets.

Trim Runaway

Trim runaway applies only to airplanes with electrically powered trim systems. Occasionally the trim will continue to move toward the full nose-up or full nose-down position after the trim switch has been released. This can lead to serious problems, particularly in the nose-up situation. The reason it is considered an emergency and not an abnormality is that, while it is usually possible to stop the runaway, it is not always possible to correct the out-of-trim condition because the trim sometimes stays stuck where stopped.

Prompt action from memory is therefore important. It is especially dangerous on airplanes that adjust the trim by changing the position of the entire horizontal stabilizer. In this case, overpowering the trim by muscle power won't completely solve the problem, because the position of the stabilizer in the out-of-trim position will limit the range of elevator effectiveness.

If an aircraft has an electric trim system, it is very important to know what the manufacturer recommends or requires in the event it runs away. The *general* procedure is to maintain control of the aircraft by overriding the trim as much as possible with muscle power, and use opposite trim on the electric trim switch to stop the runaway. Opposite trim will usually either pop the trim circuit breaker, cancel the malfunctioning switch, or neutralize the trim movement, stopping it at that position.

On certain systems you will have to hold the opposite trim until you can find the circuit breaker and pull it, or it pops itself. All of this varies from aircraft to aircraft. For your checklist, the first item will be the recommended procedure to stop the runaway (a memory item), followed by the recommended procedures to correct or disengage the faulty system.

Forced Landing

This isn't the worst thing that can happen in an airplane, but you'll never convince the passengers of that. A forced landing means simply that you have lost all power, either because of engine failure, fuel starvation, or engine shutdown. Forced landings are possible emergencies for any airplane, regardless of the number of engines. (They are much less likely with more than one engine.)

The common sense part, like trying to get to an airport or suitable emergency landing site with enough altitude to make a proper approach and landing, doesn't have to go on the checklist, and of course the checklist won't do any good if you lose all power right after takeoff. But assuming the failure occurs at altitude, a checklist can be very helpful.

EMERGENCY AND ABNORMAL PROCEDURES

The checklist should list the proper configuration and airspeed for maximum glide, along with the steps to take to configure the aircraft for maximum crash survivability. This would normally include turning the fuel off, mixture to cutoff, mags off, a note about the flaps and gear (normally flaps down for minimum ground contact speed and gear down for long and smooth terrain, gear up for short or rough), and master electrical switch off.

The checklist should also remind you to transmit a MAYDAY message, squawk 7700, and to tighten all seat belts and secure all shoulder harnesses if possible. You also want to brief the passengers on how to open the doors and exits and on what to do if you are incapacitated.

Ditching

Ditching is very similar to a forced landing situation and many of the same considerations apply. The differences are: power is available in a ditching situation, ditching is always done gear up, and to maximize the flotation time you might not want to open the main door. (But then again you might not have any choice if that is the only way out, or if the main door is the only door the raft will fit through.)

The main difference between a forced landing and a ditching is the first one though: power is available when ditching. If it is certain that you are going to run out of fuel prior to reaching land, it is better to deliberately ditch with power available than it is to let the tanks run dry and attempt a forced landing on the water. With power still available, you have the advantage of being able to keep the descent rate to a minimum in order to touch down on the water at a very flat angle. This enormously reduces the chance of flipping over or diving into the water.

Nonetheless, just prior to impact shut everything down—generators, battery, and fuel—just as you would for a forced landing. A note at the bottom of the checklist to ditch parallel to the swells and on the crest, is a helpful reminder. (Think of the swells as parallel runways.) Unless the precarious fuel situation catches you completely by surprise, you should have plenty of time to prepare the aircraft and passengers for the ditching, using the checklist to ensure completeness.

If power is not available, then this becomes a forced landing situation, only over water. In this case use the forced landing *and* the ditching checklists. The only real difference between a forced landing over water and over land is that you always keep the wheels up for a water landing and you want to make sure you can get to the raft and have a plan for getting it out the door and into it.

In airline operations, anyone doing international or overwater flying must complete a course in ditching. I did mine at TWA, and an important part of the course taught there was what they call the "wet ditching," where you actually put on the life vest and jump into the pool. We did this with the life vest fully inflated (no sweat—even if unconscious, you would probably still come up with your head well above water), partially inflated (much more difficult to keep your head above water—you probably would not float face up if you were unconscious), and uninflated prior to hitting the water (very difficult to find the inflation cords underwater).

We also found out how tiring and difficult it is to swim any distance with the vest on. The lesson to me in all of this was that you almost certainly will not be able to get your life vest on and inflated and at the right time if you have not tried it first in a pool somewhere. Just getting the vest on properly with the straps tight is an exercise in itself, and unless it is on properly, it is not going to do you any good. This is important, because without a life vest, you are not going to last long in the open seas, regardless of how good a swimmer you are.

Rapid Depressurization

For a pressurized airplane capable of operating at the higher flight levels, this is one of the major emergencies. Apparently when the door blows off a fully pressurized airplane, it sounds just like a cannon going off. The cockpit immediately fogs up, and everything loose in the airplane blows out the door, including people. (I never hang around the door or doors in a pressurized airplane, and I always keep my seatbelt on and tight if next to an emergency exit. Doors and exits do blow off and people have gone right out.)

Without supplemental oxygen, the time of useful consciousness at the higher flight levels is a matter of seconds. There literally is no time to pick up a checklist, and it might have gone out the door anyway. So the first item, for any pressurized airplane, is MASK ON and oxygen selector to 100 percent. (Not all masks have selectors, which should be left in the 100 percent oxygen position; as soon as the mask goes on this should be checked.) Getting the mask on is the most important thing—forget everything else as long as you get the mask on. If you don't get the mask on in time, you will pass out, and that's that until the airplane descends on its own through something like 20,000 feet or so. (Assuming it does descend on its own and assuming you're still alive at that time.)

There are a bunch of things to do in a rapid depressurization situation, and they should all be done quickly, so usually a memory system is developed for each airplane. In the Citation, for instance, you start by sweeping the cockpit from the left-side panel at the rear to pick up the quick donning oxygen mask and hit the switch for manual passenger mask drop, then go forward to switch the mike to the oxygen mask, then hit the front panel on the left side for the ignitors, on to the front panel center for the seat belt sign and transponder to 7700, and finally to the center console to put the throttles to idle and extend the speed brakes. This orderly sweep of the cockpit helps to remember all the necessary items. Then run the checklist to make sure you didn't skip any items.

If you are flying a pressurized airplane, it is very important that you develop a similar kind of memory system for your aircraft also—either a sweep or memory aid using easily remembered letters. The initial items for any pressurized airplane are: masks on and 100 percent oxygen selected if you have that option; communications selected to oxygen mask; passenger oxygen checked on, dropped manually as a standard back-up; no smoking. This much by itself won't solve the problem, but it will keep everyone alive and maintain communications.

EMERGENCY AND ABNORMAL PROCEDURES

Emergency Descents

Emergency oxygen is not intended to be used other than to allow enough time to maintain consciousness while descending to an altitude where oxygen is not required. You have no idea how long the emergency supply will last, nor are you assured, at the higher flight levels, that the oxygen flow from the masks will be adequate. Therefore, if the failure is catastrophic—a door or window blowing out—immediately initiate an emergency descent. If the depressurization is rapid, initiate the emergency descent then notify ATC. If the depressurization is gradual, notify ATC and request an immediate clearance to 10,000 feet. If the controller doesn't give it right away, ask again and stress the urgency—the depressurization might not stay gradual for long.

Emergency descent procedures vary from airplane to airplane also, but they all involve reducing the power to idle and creating drag, which includes adjusting the props to high rpm, rolling the airplane to the manufacturer's recommended bank angle to reduce the wing loading, and pushing the nose over to achieve a target rate of speed that depends on the recommended configuration. The idea is to get the airplane down safely and under control, but at the fastest possible rate. Whatever the manufacturer recommends should be used, and you should memorize it.

If you must initiate an emergency descent, the priorities have to be oxygen and aircraft control. But if at all possible you also want to notify ATC. This should be done both over the radio—"32 Xray, emergency descent, out of Flight Level 280 to the left"—and also by squawking 7700 on the transponder. You are going to go through many altitudes, and ATC needs to know about it.

It helps to have two pilots when you go through this drill. There is much to remember, and it is all important. I've been through it in a simulator many times. In each case I know that sometime during the training it is coming, and I have studied for it in advance; I still usually miss at least one item. I've watched other pilots go through the drill also, and very few remember every single item with any regularity because there is a lot to do in a very short period of time.

As a single pilot, your job is even harder. You have to know what the priorities are and use the checklist to cover the rest. Even though everything on the checklist is important, and even though everything should be done from memory, realism dictates that as long as you can get the mask on and start the airplane down, there will be time to get the checklist and do the other items. Don't get so excited that all you do is dive the airplane. That won't do it. And don't use "being realistic" as an excuse for not trying to do it all from memory. But, in all honesty, the only part that is *absolutely* critical is getting the oxygen mask on. The rest is critical.

If the depressurization was a result of a structural failure that might have reduced the integrity of the airframe, you might not want or be able to descend at the maximum recommended airspeed. If, for instance, a prop blade or turbine wheel came loose and ripped a hole in the cabin, you might have to baby the airplane down. If you have the mask on and start the airplane down, even slowly, you should be able to get low enough before the oxygen runs out to at least maintain consciousness. The key is to get started down.

Depressurization is not the only reason for an emergency descent. Fires, uncontrollable vibrations, and progressively worsening structural failures are others. So even if the airplane is unpressurized, you should have an emergency descent checklist modeled on the above. Omit donning the oxygen mask because you will either be low enough that you won't need it or will already have it on.

If the manual does not describe an emergency descent procedure, try this one: power to idle, prop forward, bank 45 degrees, **either** gear *down*, target speed of maximum gear extended speed; **or** gear *up*, top of the yellow airspeed arc in smooth air, top of the green arc in less than smooth air. Experiment to see which configuration provides the maximum rate of descent. (Pick a good day to try this and make sure to clear underneath first.) You don't have to bring the power all the way back to idle for comparison purposes, but if you really want to know the actual maximum rate of descent, the power will have to be reduced to idle. This isn't the best way to treat an engine, but if you have the engine temperature stabilized as cool as possible prior to reducing the power, the stress should be minimal.

Failure of All Generators/Alternators

This "failure" means failure of *the* generator on most single-engine airplanes and failure of both generators on most twin engine airplanes. This is a marginal emergency; it could almost be considered an abnormal situation, but with everything turned on and all generating power gone, the battery in most airplanes will usually only last a few minutes, so time is important. Probably the only real memory item is to remember to use the checklist.

The checklist should include whatever information is necessary to verify that the generators are indeed off line, and the procedure for attempting to reset them. If the resets fail, a checklist should then remind you to reduce the electrical load as much as possible. A list of nonessential items is helpful here: pitot heat (takes a huge amount of juice), secondary navs and coms, all the lights, and the turn-and-bank. (For items not having switches, pull the controlling circuit breaker.)

Any unnecessary equipment—which is normally everything except one nav and one com, the transponder, and any electrically operated primary gyro instruments—should be turned off, at least initially. If the juice lasts until final approach, you can always turn the pitot heat, landing lights, flap and gear controls, back on for the landing. If power doesn't last, you will have to lower the gear manually and remember that the gear indicator lights won't work. This should be noted on the checklist, along with a reminder to notify ATC of the problem and request a vector or descent to VFR conditions.

A note listing all the items that *will* operate without electrical power is reassuring at the bottom of this checklist also. Possibilities include the engine, compass, altimeter, airspeed indicator, vacuum-operated gyros, vertical speed indicator, windup clock, and one or two of the engine instruments. The easiest way to find out what works and what doesn't—a very useful lesson—is to fire the airplane up on the ground and then turn the

EMERGENCY AND ABNORMAL PROCEDURES

master and generators off. Whatever is still working is what you have left after the battery runs down.

Spins

The recommended procedure to recover from a spin, for your aircraft, should be on the emergency checklist, not because you will have time to dig it out and use it, but because it will help to fix the procedure in your head and be useful for review purposes.

The classical spin recovery procedure is: opposite rudder, stick neutral, power idle; when the rotation stops, rudder neutral and pull out gently from the dive. Airplanes respond differently to spins though—it depends a lot on how the airplane was designed, so do whatever the manufacturer recommends. If the manufacturer prohibits spins, believe it. If spins are prohibited it means either that the airplane was never fully spun in the test program—and nobody really knows what it will do in a fully developed spin—or that it is possible, with certain weight-and-balance situations, to enter a spin from which recovery is impossible.

ABNORMAL SITUATION CHECKLISTS

So much for emergency procedures. Abnormal situations do not require an immediate, memorized response, so the more situations you can think of and prepare for ahead of time with a checklist, the better. The checklist serves two important functions; one is to help troubleshoot the situation and the other is to provide the best and most complete response. No matter how well you might know your aircraft systems, there is no reason not to have a checklist to at least back you up. The smart thing to do when an abnormal situation arises is to go straight to the checklist and let the checklist systematically move through your troubleshooting analysis, if appropriate, and solution. This is the error-free way to do it.

Here are some typical abnormal situations: generator failure (when more than one generator is available), starting problems, loss of hydraulic pressure, flaps stuck or split, precautionary shutdown, engine inoperative, no nosewheel steering, brake failure, pitot heat inop, fuel pump inop, low fuel pressure, low oil pressure, high oil temperature, high cylinder head temperature, gear problems, overpressurization, heater inop, door not latched, vacuum pump failure, static system clogged, and induction icing.

I'm not going to go into the same detail with each of these that I did with the emergency situations, but I do want to give you an example of a situation that requires troubleshooting prior to being able to take the proper corrective action, and demonstrate what that kind of checklist might look like. I also want to point out some general principles to follow when preparing abnormal procedures checklists, and I will make some comments about certain situations. With that you should be able to do a pretty good job of making up your own abnormal procedures checklists.

Abnormal Checklists

Generator Failure

Figure 9-2 is a sample abnormal checklist for generator failure—an electrical problem. People always have trouble with electrical systems, myself included, and this is one of the best situations to use a checklist. A generator (or alternator) can fail, or appear to fail, in several different ways. Merely because a warning light comes on it does not necessarily mean that all generating power has been permanently lost.

Electrical systems vary enormously; prepare a checklist based on the manufacturer's recommendations for your particular airplane. This is an example of what you want to create; something that helps you troubleshoot the problem, and then directs you to the proper corrective action depending on the outcome.

GENERATOR FAILURE
Generator warning light ON.

1. Check circuit breaker.
2. Check generator switch ON.
3. Check voltmeter.
   a. Zero voltage.

      Attempt reset—Switch to OFF, then ON.
      If light stays off, continue to use generator.
      If light comes back ON, select
      generator OFF, proceed to Step 4.

   b. Voltage normal
      Reset not possible. Select generator OFF.
      Proceed to Step 4.

   c. Voltage high.
      If higher than 30 volts select OFF.
      Proceed to Step 4.
      If fewer than 30 volts, continue to use generator.
      Monitor voltmeter.
      Report to maintenance.

4. Generator selected OFF.
   a. Reduced load on remaining generator to 100 amp.
   b. Review emergency checklist for double generator failure.

Fig. 9-2. *This is a typical abnormal procedures checklist for generator failure. At item number three, the checklist branches into three troubleshooting possibilities, labeled a, b, and c, with a recommended action for each condition. There are no memory items.*

EMERGENCY AND ABNORMAL PROCEDURES

I don't believe in troubleshooting for the sake of troubleshooting. Leave that to the A&P mechanics. If you have an idea what the problem is, fine, but the main idea is to solve the problem, not identify its cause. If, for instance, you were using the checklist in FIG. 9-2 after a generator warning light had come on, and you discovered that the voltage read zero, and you attempted to reset the generator but the light came back on, don't worry about trying to figure out if it was the generator that failed, or the voltage regulator, or the gauge, or something else—just turn the switch off like the checklist says. You've done all you can. Let maintenance figure out what needs fixing. If you spend the rest of the trip trying to figure it out, you'll probably mess something else up.

For any system for which a circuit breaker is a part of the system, the first step on the checklist should be to check the circuit breaker. You'd be surprised how many problems can be solved by pushing a circuit breaker back in. If it pops again, wait a minute or so for the breaker to cool and try it one more time. The general rule for circuit breakers is that you get two tries per circuit breaker, after that, leave it out and proceed with the rest of the checklist.

If a system has a switch, checking that the switch is ON should be part of the checklist. Switches quite often get knocked OFF, and it's very frustrating trying to troubleshoot a system that is turned off. You might say these two checks are obvious, but I can't tell you how many times a checklist has helped me catch a popped circuit breaker or switch knocked off.

Loss of Hydraulic Pressure

Most general aviation aircraft have simple hydraulic systems, so the loss of hydraulic pressure is not nearly so serious for them as it is for larger, transport aircraft where the flight controls are generally hydraulically boosted. In fact, in most cases the only thing the hydraulic system operates is the gear, and there is always a manual backup for all gear systems. In any case, the loss of hydraulic pressure is not in itself a problem—the *systems* you lose are a problem. So the main point of this abnormal checklist is to note which systems are affected and to refer you to those checklists: "See Landing Gear—manual extension checklist."

Flaps Stuck or Split

The problem with the flaps being stuck in an extended position is that airspeed will be limited and cruise performance will be reduced. The checklist should have a listing of what the speed restrictions are for each increment of flaps, and a reminder of the reduction in altitude and range capability, with the suggestion that a return to the departure airport or takeoff alternate be considered.

Flaps split, where one side has a greater flap extension than the other, is a serious problem. Assuming the airplane is within lateral balance limits (you have observed all restrictions on fuel balance from side to side), you should be able to control the airplane even with flaps full down on one side and all the way up on the other. The checklist should cover whatever procedures the manufacturer recommends for a particular aircraft to try to get the flaps unsplit—this is no way to try to fly an airplane. In most cases, if the flaps cannot be unsplit, it is better to go back to the last setting so they are at least the same on each side, even if that means extending the flaps further on climbout or reducing them on approach. Then refer to the "Flaps Stuck" checklist.

Precautionary Shutdown

It helps on this checklist to have a list of possible reasons for shutting an engine down. This will help you make the decision to go ahead and shut one down, if that is indicated, and slow you down from unnecessarily shutting one down for various false alarms. Good reasons for shutting an engine down are: rapidly rising oil or cylinder head temperatures with the oil pressure dropping; both oil and cylinder head temps rising rapidly; any temperature over redline that cannot be reduced and is confirmed as not a gauge problem; bad or worsening vibration; uncontrollable prop speed. The question here isn't saving the engine because if you have one of these problems, the engine is probably shot already. Rather, it is a question of whether the risk of the situation getting catastrophically worse and causing structural damage to the airplane is greater than the risk of operating without the engine.

Once you have determined that a precautionary shutdown is in order, the checklist should lead you through an orderly, step-by-step procedure for shutting the engine down and cleaning up the airplane with a referral to the "Engine Inoperative" checklist at the end.

Engine Inoperative

A reminder of crossfeeding procedures is a good idea here, as is a note on the single-engine service ceiling at various weights if that information is available, and a reminder of the best single-engine rate-of-climb speed.

Pitot Heat Inop

If you don't have a warning light installed for pitot heat inop, you probably won't know when the pitot heat is inoperative until the airspeed goes nuts. Erroneous airspeed indications can be very hard to identify. A warning light is a great idea, but if your airplane doesn't have one, be very skeptical of airspeed indications that don't seem quite right.

If you aren't sure whether the pitot heat is working or not, turn it off and look for a drop in load, and then turn it back on and look for a rise. If nothing happens, the pitot heat is inoperative. If it is inoperative, disregard all airspeed indications in icing conditions.

An old instrument flying adage is "Power plus attitude equals performance." This means that any given attitude and power setting will result in a given airspeed and rate-of-climb (positive, negative, or zero). If you know from experience, for instance, that a level attitude and 1,800 rpm results in an approach speed of about 80 knots and a descent rate of about 500 fpm, then you don't need an airspeed indicator, do you? As long as the attitude is set to level and the power is set to 1,800 rpm, the airplane will settle at 80 knots and 500 fpm (or very close to it), with or without an airspeed indicator. This is where "knowing your airplane" pays off. This is also where a note to check the switch and circuit breaker pays off.

Gear Problems

This is an important checklist, and the more complicated the system, the more important it is. You particularly want to know what all the various possible light combinations mean if the system has both an unlocked or in-transit light, and gear down-and-locked lights. You can get some pretty confusing indications. Does a warning light simply mean not locked up, something in between, or definitely not locked down? What does it mean if you get both a warning light *and* three green lights? On some airplanes, this just means there is a problem with the in-transit circuit, but the gear is definitely down and locked. On others, anything other than just "three green" implies a locking problem. What precautions are recommended by the manufacturer whenever there is any abnormal gear indication? What is the step-by-step manual extension procedure? Finally, what is the recommended procedure for a gear-up landing?

Heater Inop

An airplane will cool off fast when the heater quits, particularly with an outside temperature of 20 or 30 below. In fact, at these temperatures, it doesn't take long to get so cold you literally can't fly the airplane. If there is any way to get a failed heater started again you want to know what it is, and you want it on the checklist so you don't have to waste time looking it up in the book.

Vacuum Pump Failure

What instruments and systems are vacuum operated? Make up a list, with backup instruments, so that when it happens you don't have to experiment or guess.

Static System Clogged

A clogged static system results in erroneous airspeed, altimeter and vertical rate indications. The checklist should remind you of the alternate or emergency static source location. It should also either list or direct you to the airspeed compensation chart for the alternate source, or advise you of the amount of error that can be expected.

PROFESSIONALISM

It is a lot of work to make up an emergency and abnormal checklist for each situation, but it is extremely worthwhile, and I can't think of a better way to truly master the systems in an airplane. A checklist-oriented solution to problem solving is the professional way of dealing with emergency and abnormal situations, and, in my opinion, it is one of the main reasons flying on a professionally flown aircraft is so much safer, statistically, than on a nonprofessionally flown aircraft.

"General aviation," which includes corporate aviation but which is predominately private aviation, has an average accident rate per flight hour about 30 times greater than that for the scheduled airlines, according to figures for the last several years from the National Transportation Safety Board.

The airlines certainly aren't safer merely because their pilots are getting paid—the airplane couldn't care less whether the guy manipulating the controls gets a paycheck or not—and professionals aren't supermen or geniuses either. But they *do* have checklists to help them out when things go wrong, and they *have* been trained to use them. There is no reason why you can't do the same thing.

If nothing else, I hope this chapter has caused you to review and rethink your attitude toward emergency and abnormal situations. The professional doesn't just hope nothing ever goes wrong, and he doesn't just assume that if it does his experience and general knowledge of the systems will be enough in itself to solve the problem. Most professionals are smart enough to know they are not that smart.

One of the more obvious but frequently overlooked differences between a professional pilot and an amateur pilot is that the professional flies a lot more. He flies enough, more so, that over the years a lot of these emergency and abnormal situations actually do happen to him (or to her). He only has to scare himself once or twice by responding in an erratic and non-systematic way to learn that things *will* go wrong and that ingenuity and his superior nerves alone won't always save the day.

The ground school instructor I had when I was studying for my ATP written told a story about two different times he, as an Air Force navigator on C-130s, participated in actual precautionary engine shutdowns. The first time, as soon as a warning light came on, everybody in the cockpit started trying to solve the problem at the same time, only everyone had a different idea of how to do it. Somehow they managed to shut the wrong engine down, had to get it going again, and finally got the desired one shutdown before anyone thought to use the checklist, and then discovered that they had done it wrong and had to retrace their steps and do half of it over again.

The second time, with another crew, the engineer announced the warning light, the plane commander directed the copilot to read the engine shutdown checklist, and everybody in the cockpit (including the navigator, as per procedure) verified that the plane commander had his hand on the proper control for the proper engine prior to his moving it. Within a matter of seconds the engine was shutdown, the clean-up and crossfeed items were attended to, and everybody went back to drinking coffee and telling war stories.

That's the way it's supposed to be done, but to do it takes two things: a checklist and self-control. A good place to start is with the checklist.

10
Proficiency

PILOTS TALK A LOT ABOUT PROFICIENCY. THE WORD GETS USED IN A VARIETY OF WAYS, but usually "proficiency" means "maintaining basic flight proficiency." That's jumping ahead a little though, because before proficiency can be maintained, it has to be acquired. Proficiency thus has two parts, the first being the *acquisition* of the knowledge and skills essential to safe flight, and the other the *maintenance* of that knowledge and those skills through training and practice.

The first chapter (The Basics) tried to cover the areas that a serious pilot should be familiar with, and described the standards he or she must achieve to function smoothly and properly in the aviation transportation system. That description was, however, necessarily somewhat simplified. It was necessary, at that point, to list and describe basic flying skills and their allowable tolerances without too much further elaboration. There was no practical reason to go into any more detail until those skills and standards were established. Chapter 1 described the basic skills and this chapter describes the manner and method of achieving and maintaining those skills. That is, in flying shorthand, "proficiency."

THE PROCESS

While it was necessary at some point to list the skills and standards necessary to achieve minimum levels of proficiency, that doesn't mean that "proficiency" is simply a matter of achieving a certain minimum level of skill and then maintaining that level. Proficiency is more complicated than that.

PROFICIENCY

Think back to an experience in your life that was particularly meaningful and satisfying but very demanding—some activity you became very skilled at and were very proud of, like an experience on a winning team, or some special challenge or training in the military, or a big project at school or at home, or even some aspect of your work that you became especially skilled at. Perhaps you will see what I mean by "proficiency" not being a simple matter.

If you reflect back on this accomplishment, you will probably remember that you were very proud of yourself, and when all was said and done you were very glad you had undertaken this project, but that there had been moments of serious doubt and discouragement along the way. In addition, at some point after you had achieved your goal you probably realized that you had achieved a higher level of skill than the minimum needed to merely get the job done. But you probably also remember that as soon as this particular experience was over—the team broke up, or you left the military, or finished the project, or your job changed—that the new skills deteriorated almost instantly from the peak you had achieved.

As time went by, it probably seemed that *all* the skill and knowledge—which had taken so much effort to acquire—disappeared and there wasn't much you could do to hang on to it. Even now, though, you are probably quite certain that you could go back to that activity, and, with a little review, be just as good as you ever were, and maybe even be just a little better.

Specific facts and essential details tend to be forgotten very quickly once we quit reinforcing them with daily use. But the important parts of an accomplishment—the physical skills, the general principles, the lessons of experience—stay with us forever, and the specific facts and essential details come back quickly.

Flying well is one of these demanding but satisfying experiences. If you have been flying for a while, you know that achieving true proficiency is an awful lot of work and is frequently very discouraging, but, having achieved it, you know that there is a lot to be proud of and you would gladly do it again. You know that after achieving real proficiency in aviation, that your level of accomplishment is actually greater than that needed to merely get the job done, but you also know that it is the extra level of accomplishment that gets you through the tight situations.

You know that peak levels of skill can only be achieved with great effort, and maintained at a peak for very short periods of time—that you have to fly almost all the time to maintain a peak, and even then the nonroutine aspects fade from inevitable neglect—but you also know that, having achieved a peak of proficiency, you can tolerate a little bit of slippage. Finally, you know that if you were to quit flying that it wouldn't be too long before you would forget all the essential facts and figures, but you also feel that with a little effort they could be reacquired very quickly.

This process we go through whenever we achieve a high level of skill and competence in some area, which I have tried to break down into its component parts, is really what "proficiency" is all about. I want to use this process as the structure for talking about the specifics of proficiency—what, as a pilot, do you *do* about proficiency?

ACQUISITION

As it is with any difficult but worthwhile activity, basic proficiency doesn't come easily, nor in the case of aviation, does it come cheaply. There is a lot of work involved before anyone can achieve minimum levels of competence. It takes time and study and practice before you can consistently keep the altitudes within 100 feet and the headings within a couple of degrees and the speeds within a few knots and before you really know an airplane's systems and don't have to fumble for switches or worry about forgetting little things or having to look everything up.

There are no short cuts to acquiring proficiency. It is slow, it is discouraging at times, and it is expensive. But proficiency, in anything, has to be built on a foundation of fundamental competence. You have to have achieved basic mastery of your instrument before you can sit down with the orchestra.

MARGIN OF SAFETY

When you have truly mastered the fundamentals, your level of skill will be much better than it "needs" to be. When you are really comfortable with your airplane and with the instrument flying system, and you look back on what you have accomplished, you will find that you have achieved much more than the minimum. Minimum levels of competence seem unreasonably high when just starting out, which is good, because it forces you to work hard to achieve them, and, in working hard to just reach the minimum, you inevitably do more than necessary.

It is very important that this "extra" level of skill beyond the minimum be achieved because that is where the safety factor comes from. If you have only just mastered the basics to minimum standards, there is no way to allow for extraordinary or demanding circumstances. This is just common sense again. You have to achieve at least minimum standards, but to be sure of never dropping below those minimums you have to be better than just minimum—there has to be a cushion to fall back on.

In practical terms this extra margin is not anything you have to consciously worry about. You don't, for instance, have to raise your personal minimum standards to be sure of staying within the legal minimums. The legal minimums alone are demanding enough. If you can meet those standards (plus or minus 100 feet and 5 degrees and so on), you will already have the necessary skills to do better than that—the highest hurdle will already have been jumped.

You will know that you have gone beyond the minimum when you don't have to struggle or even think about holding altitude and heading and speed. One day you will be flying along and will realize that without working at it, everything is within limits and there is no reason why it can't always be that way. At this point you have acquired proficiency.

CHECKRIDES

Now the critical area comes into focus: *maintaining* proficiency. One deficiency in Part 91 is the lack of a requirement for regular recurrency training and checkrides. Any

Proficiency

pilot-in-command operating under Parts 135 or 121 (the regs that cover air taxi operators, the commuter airlines, and the major airlines) has to have a proficiency check every six months. In addition, any pilot-in-command of a Part 25 (Transport Category) aircraft has to have a proficiency check every 12 months, regardless of what part he operates under. (This is at least a partial admission, on the FAA's part, that there is also a need for a requirement for regular checkrides under Part 91.) The checkrides basically cover the normal, abnormal, and emergency operation of the systems, the usual airwork, single-engine procedures, and a series of approaches.

Proficiency checks normally aren't quite as tough as an initial checkride for a type rating (there is usually no oral, for one thing), but they can be, and at any rate they don't miss by much. Few professional pilots really get a kick out of taking checkrides, but all are in favor of them, at least in principle.

I know of very, very few instances where pilots have "busted" a check. It does happen and it can happen to any pilot at any time if he has an especially bad day or the examiner is determined, for some perverse reason, to "bust" him, but that is rare. The point of checkrides is not to weed out bad pilots. That might be the official justification, but that is not the real reason. The real reason for checkrides is to force pilots to regularly bring their levels of competence up to a peak.

No matter how many checkrides you take or how experienced you are, the pressure is always on when a checkride is due. That pressure is what forces a pilot to open the books, review the abnormal and emergency procedures, pay attention during the training sessions, and generally do whatever is necessary to bring flying abilities up to a peak to avoid embarrassment on the checkride.

RECURRENCY TRAINING

It's a tough system, but it works. Every time I go to recurrency training I ask myself why I didn't take up something simple like brain surgery instead of aviation. I always get through it and I am always a much better pilot coming out than going in. No matter how lazy I am between checkrides (I always mean to get my manuals out a lot more often than I do) and no matter how undemanding or routine the flying is between checks, my overall flying won't deteriorate too badly, because I am going to be "back in the wringer" in the not too distant future.

The result, for most professional pilots flying Transport Category airplanes, is that while their proficiency varies from checkride to checkride, it starts at an elevated level as a result of the fairly intense initial training and testing for the type rating, and it never drops very far below that level because of the frequent checks after that. This is true "maintenance of proficiency."

Part 91 does not have a requirement for either regular recurrency training or for checkrides (except for pilots of Transport Category airplanes). The requirement for a biennial flight review is a step in the right direction, but the interval is too long and it isn't a checkride—it's a "review" (as opposed to a "test"), which means there isn't the same incentive to prepare as there is for a checkride.

MAINTAINING PROFICIENCY

Assuming you have achieved basic mastery of the airplane, including its systems and procedures, and can operate comfortably within the limits of the instrument flying system, the questions, become "How do you maintain that basic proficiency," and two, "How do you stay current in your emergency and abnormal procedures?"

Recall an earlier thought that the best answer to the question is regular recurrency training with checkrides to remain honest. Unfortunately, formal recurrency training for pilots of personal aircraft is limited at this point. It does exist for some types and the major training organizations are working with the aircraft manufacturers to add aircraft to the list of aircraft ground school/simulator programs all the time. School is the easiest and best way possible to maintain proficiency. Sign up right now to go at least once a year, twice is better. Write it off as another fixed cost of aircraft ownership.

SIMULATOR TRAINING

Simulator training is the best way to maintain proficiency because the training is thorough and, in an ironic way, simulator training is also much more realistic than flight training. Simulator training covers every conceivable situation that can be simulated, practiced, and analyzed; it is more realistic because there are many abnormal and emergency situations that cannot be realistically simulated in an aircraft, either for safety reasons, or simply as a matter of practicality: engine fires, gear failure, hydraulic failures, trim runaways, total electrical failures, dead stick landings, brake failures, prop overspeed, fuel leaks, and more. The reason the military, the airlines, and most corporate flight departments believe in using simulators for initial and recurrent training is not because simulators are cheaper than airplanes for training—in some cases they are, but that's a side benefit—it is because simulation is better than flying for *training*.

If your airplane is not one for which regular recurrency training is available, be patient. In the meantime, do what you can, assuming you have succeeded in acquiring at least basic proficiency, to maintain proficiency.

FREQUENCY

Probably the first and most important single thing a pilot can do to maintain proficiency is to fly often. There simply is no substitute for regular flying. When I haven't flown for a couple of weeks, I can tell when I get back that I am not quite as proficient as I was before. The switches and knobs don't fall to hand like they should, the checklist doesn't go as smoothly, my eyes have trouble finding things, important numbers don't leap into my head, and so on. Everything returns quickly, but even after a couple of weeks away, I can tell. It is hard to say what "regular" flying is, and too much flying can sometimes be a problem too, but I would guess that something like three to four flights a week is optimum.

PROFICIENCY

Certainly you probably need to fly several times a month to even hope to maintain basic proficiency. I don't know how many "several" is, but it is more than two or three. This may mean having to take the airplane out of the hangar and fly it even though you don't have anywhere to go, but there are things you can do to get more out of usual flying.

A "flight" is a cycle: start-up, taxi, takeoff, climb, cruise, descent, approach, landing, taxi, shut-down. You have to go through this cycle for every flight, regardless of the length. As far as proficiency goes, the number of *cycles* is much more important than the number of *hours*. Maintain proficiency with frequent flights. From the viewpoint of maintaining proficiency, two 250-mile legs are better than one 500-mile leg. The time spent sitting at altitude with the autopilot on really doesn't do much for proficiency, but an extra takeoff, descent, and landing does. Two legs may be less efficient than one, but that is a small price to pay to maintain proficiency.

If you own the airplane, you have an advantage over the renter-pilot in that the only significant operational expense—variable cost—is fuel. (If your airplane is maintained on a system based on hours-in-service, and it should be, as opposed to a simple annual inspection, there is an additional expense for maintenance each time you fly, but the hourly cost is relatively small. Depreciation is also a big expense, but depreciation is generally a direct reflection of the age of the airplane and doesn't increase much with time on the airframe. In fact, fewer than 300 hours per year might be a liability. Each hour flown also puts an engine that much closer to overhaul time, but most engines are overhauled due to lack of activity, which results in corrosion and premature wear, not because they have run out of time. In this case more frequent flying might save a little money. Therefore, as a practical matter, the only significant operational expense for the owner/pilot is fuel.)

If the airplane holds four, but you have six who would like to go, make two trips. If a partner lives 25 miles away and needs to go with you on a trip, don't make him drive to your airport, fly over to his and pick him up. It might not make perfect sense economically and it might not even save any time, however you have the airplane and a good part of the cost of ownership must be paid whether you fly or not, and the more you fly the airplane the better. Besides, any valid reason to fly is more economical than going up merely to practice. If you find you can't afford to be this proficient, that's another question—you might have too much airplane. But you certainly can't afford to not be proficient.

The pilot who rents has to look at that big hourly fee every time he thinks about flying somewhere. Very often that's all it takes to kill the idea, and he finds himself flying less and less, and pretty soon his proficiency is gone. This is the wrong way to look at it. It does cost a lot to rent, but in most cases renting still doesn't cost as much as the total cost of owning an airplane. The breakeven point for owning over renting usually doesn't come until after 300 to 400 hours of operation a year. So the plus side for the renter-pilot is to take all the money he saves not owning an airplane and use it for rentals. The pilot who rents needs to fly just as much and just as often as the owner-pilot does. The airplane doesn't know or care whether you have an equity interest in it or not—it simply insists on being flown by a proficient pilot.

Systems

Flying as much as possible is the single best thing to maintain proficiency. But you know, and I know, that no matter how much you fly it won't be enough. Flying for the nonprofessional always has to be squeezed in between job and family, and it is usually a tight fit. It is like going to the doctor for a physical—you never get enough exercise to satisfy him. But the doctor also knows that if he doesn't nag you a little bit you'll exercise even less. So I'm nagging you a little bit.

Fly.

Fly every chance you get.

Look for reasons to fly.

If you have to, fly even when there isn't any reason.

And to get the most out of the flying you are able to do, plan each flight carefully, even the short ones. You get much more out of one flight done properly than five done poorly or out of habit.

Make each flight count, and fly as much as you can.

SYSTEMS

Flying all the time is the key to maintaining proficiency in the basic skills, but there is another part to proficiency and that is proficiency with the aircraft systems. Chapter 9 covers a variety of abnormal and emergency situations and reaches a conclusion that a checklist is the best tool for dealing with these problems. But having a checklist and being able to use it are two different things.

Proficiency means being able to read the checklist and do what it says while flying the airplane during an emergency or abnormal condition. Proficiency also means having a thorough understanding of the systems so you can deal with the situations that don't fit the checklist. Regular flying, unfortunately, doesn't do a thing for these situations, because abnormal and emergency situations are not a part of regular flying.

One thing you can do to maintain proficiency in abnormal and emergency procedures is get the operating handbook and flight manual out and read them, or at least parts of them, with some regularity. Every time I get my manuals out I find things I had either forgotten about, never knew, or remembered incorrectly. (Remembering incorrectly is probably more dangerous than not knowing at all.)

One thing we learn as we get older is that self-discipline is the hardest form of discipline; people actually do us favors when they require our service and don't do us favors when they allow us to discipline ourselves. The FAA requires me to study manuals at least every 12 months for each aircraft I am typed-rated and current in, which is truly doing me a favor.

I might not "get around to it" if they didn't. Probably the FAA would actually be doing you a favor if it required all Part 91 pilots to study the manuals on a 12-month basis, instead of only pilots flying Part 25 aircraft. But that is not the case, so you're going to have to discipline yourself.

If the manual pages are starting to stick together, you aren't reading them enough.

PROFICIENCY

PROCEDURES TRAINING

So now you're flying all the time and doing a good job of maintaining basic flying proficiency and you read the manuals regularly and know the systems and abnormal and emergency procedures pretty well. What else can you do? One thing is to sit in the airplane with the abnormal and emergency checklists and go over each procedure. Pick an emergency or abnormal area at random, drill yourself on the memory items, and then go to the checklist to make sure you understand it and can find everything listed quickly and accurately.

It is a good idea to actually move the controls, turn the knobs, throw the switches, and so on. This sometimes reveals things you thought you knew but don't—it is awfully easy to assume you know where everything is, when in fact you don't, or do but can't figure out how to get the door open or the knob to turn, or whatever. In addition, actually doing it helps fix the procedure in your memory.

If the FBO has a power cart that supplies the correct voltage to your airplane, and the airplane has an external power receptacle, hook up the cart to power all aircraft systems (except the engines of course) without draining the battery. This turns the airplane into a procedures trainer, something normally found only at training centers.

Several different types of procedures trainers exist. The simplest kind is nothing more than a full-size picture of the cockpit layout. As you go through the procedures you point and touch the appropriate switch or control on the picture. These are very often used in formal ground schools for initial training.

The next step is a model cockpit, with real switches and levers to move, but no power—nothing happens. The next step is to power the instruments so that the lights and indicators go on and off as various switches are thrown, and the last step is to hook the procedures trainer up to a computer so the gauges and indicators respond to the movement of the switches and levers.

Hooking an airplane to a power source creates a third level of procedures trainer, which isn't bad at all. You can go through all procedures and, as you throw the switches and levers in response to the checklist, in many cases something will *happen*—lights will go out or come on, electrical loads will rise and fall, warning flags will appear, and so on. (Don't move the gear switch. I know that probably goes without saying, but I don't want a pilot to say I told him to try to raise the gear on the ground.) It might sound dumb to think about sitting in the airplane alone playing "pilot," but if it is dumb, the word hasn't gotten to professional aviation yet; go to any airline or corporate training center and see pilots sitting in little rooms talking to themselves, playing with knobs and switches.

FLIGHT TRAINING

All of this great, but there is still one part missing, and that is getting into the airplane (assuming again, that you don't have access to a simulator) and actually going through the procedures. I think every pilot who doesn't have access to simulator training, regardless of the type of airplane he flies, should go up at least once a year with an instructor who has a lot of experience with his particular airplane for a workout, even if his particular airplane is a Champ.

Single-Engine

Obviously, in the case of a Champ you wouldn't spend too much time on emergency procedures because there aren't very many in a Champ. There are some though because a Champ can have an engine fire, too. But you can do some airwork—a good exercise is maneuvering from minimum controllable airspeeds through partial stalls to full, accelerated stalls. Work on the maneuvers requiring precision, like short-field and crosswind takeoffs and landings.

A good instructor will also throw in a couple of unexpected engine failures and will know of a runway somewhere in the local area that is marked to simulate a short field. (It is nice to know how short a field you can get one into if you have to.) All of this, to one degree or another, applies to all single-engine operations. For instrument-rated pilots, the training should also include a review of instrument approach procedures and at least one hold.

Multiengine

Multiengine pilots should emphasize instrument work, single-engine operations, and aircraft systems. (Aircraft systems are invariably fairly complex in multiengine aircraft.) This doesn't have to take all day because many elements can be combined. For instance, on one instrument approach the instructor can simulate a gear problem—the gear never acts up on good days—and on another approach, he can "fail" an engine.

I know this kind of drill is a lot of work and not too much fun, but, the fact is, you asked for it when you decided to fly a multiengine airplane. You get a lot of performance and utility out of a sophisticated twin, but the price paid for that utility is the requirement to fly it safely. Pilot capabilities have to match the airplanes. The more sophisticated the airplane, the more flying you have to do, the more work it is, and the more self-discipline is necessary to stay with it. For the part-time pilot without access to simulator training, maintaining proficiency in a sophisticated multiengine airplane can be a real challenge.

In general, to maintain proficiency you have to fly a lot, study manuals regularly, sit in the airplane once in awhile and practice emergency and abnormal procedures, and fly with somebody who is on top of the airplane and who can provide a good workout in it at least every 12 months. (If you fly less than four or five legs per month, then you ought to reduce this interval to every six months.)

If you do all of these things, I think you can be confident that you are doing all you reasonably can to maintain proficiency. You will also be doing more than 99 percent of all other nonprofessional pilots, and you deserve to take real pride in the ability to maintain proficiency in a demanding skill above and beyond your primary profession or occupation.

LOSS OF PROFICIENCY

The *loss* of basic proficiency is a distressing but sometimes unavoidable occurrence. Partner problems, special assignments, transfers, financial setbacks, kids coming along, kids getting braces, kids going to college, kids getting married—many things can lead to long periods of flying inactivity and it doesn't take very long to become very rusty. The first thing to go is the ability to recall important numbers, followed by imprecision with instrument procedures. Then proficiency with, and knowledge of, the aircraft systems starts to fade away into a vague blur, and after a year or so it's hard to remember which system operated the gear or the flaps, much less how. The mind seems to be programmed to dump information we aren't using in order to save memory capacity for new areas, even when we don't want it to.

The good news is that memory returns very quickly, especially the physical aspects, so don't worry about it very much. When I was working as a flight instructor, someone would occasionally come in who hadn't flown for 25 or 30 years. It was incredible how fast the basic ability to fly the airplane came back. The knowledge part was pretty much long gone, but the physical manipulation of the airplane came back very quickly, and the rest didn't take nearly as long to "retrieve" as you might think. Our long term memory might dump a lot of stuff when we quit using it, but it seems to leave an imprint of some sort, so that it can be filled in again very quickly.

If you have to quit flying temporarily, accept the fact that proficiency will deteriorate very rapidly, and that there isn't much of anything you can do about it. You can attempt to stay proficient by reviewing manuals, and reading the flying magazines, and hanging around the airport, but all you're doing is delaying the inevitable at best, and you might be deluding yourself into thinking you are still proficient. You're better off putting it out of your head and not worrying about it. Don't plague yourself with guilt or resolutions you can't keep. If you want to fly again someday, use your energy and time to do whatever is necessary to fly again, rather than struggling to try to maintain an illusion of proficiency.

The knowledge that the loss of proficiency is not irreversible or permanent should provide the peace of mind to pursue other priorities until flying can re-enter the picture. This is true whether your break from flying is six months or six years. The bulk of the loss occurs very quickly—within a year—and any additional time after that results in very small increments of additional loss. In fact, the biggest problem after a year away is not "rustiness" (the further deterioration of proficiency), but obsolescence—things change and you have that much more catching up to do when you come back.

REGAINING PROFICIENCY

The process of regaining proficiency is exactly the same process as acquiring proficiency in the first place, only greatly accelerated. The parts that you have not forgotten will be recognized immediately and skimmed over.

A few years back I decided to try my hand at gliders again, after 10 years. I told the instructor that I had previous instruction and experience with gliders, but that I didn't really know how much I remembered, so I wanted him to assume that I was a new student and take it from there. In fact I was surprised how much I did remember, but I was also surprised at what I had forgotten, like how to use my feet for coordination. (Jets have short wings that don't generate much adverse yaw in a turn, and the yaw dampener takes care of what little rudder is needed anyway, so my feet were very lazy.) Starting from scratch was the right thing to do and put everything in the right order, the instruction had continuity, and we didn't skip anything. Everything came back very quickly.

If you decide to get back into flying after some time away, start with the regulations, then basic VFR airwork, then concentrate on the airplane and its systems, move to IFR procedures, and finally emergency and abnormal procedures.

If you've been away less than a year, this process won't amount to much more than what you should do every six or 12 months anyway. If you've been away longer than that it might take several sessions, but the important point is, you don't have to learn to fly all over again. It does take a little work, but a lot less work than it takes to worry or fuss over it for a year or two trying to stay "sort of" proficient.

At any rate you are being honest with yourself this way: when you quit flying you know you lose proficiency; when you start again you know you need a little help to get it back. There can't be anything half-way about proficiency; a little bit of proficiency can be a dangerous thing.

EXCUSES, EGOS, AND REWARDS

Proficiency will always be a problem for part-time pilots. Almost all of their instruction and flying has to come out of their own pockets, and they usually aren't able to fly as much as they really should to maintain basic flying skills, much less to maintain proficiency with emergency and abnormal procedures. But the biggest problem is getting part-time pilots to understand that neither the airspace system nor the airplane will make allowances for them—there is no handicap system in aviation. The system is not very tolerant—it demands instrument proficiency, and airplanes are not very forgiving, they demand competence. Neither gives any points for extenuating or mitigating circumstances. Serious flying is like playing hardball with the big kids: if you want to play with them you can, but don't expect them to make any allowances for you, and if you get hurt, don't expect too much sympathy because they didn't ask you to play in the first place.

There isn't any way to sugar-coat the importance of proficiency. The lack of basic proficiency among pilots of personal aircraft is a serious problem and general aviation has the accident record to prove it. There is no reason a part-time pilot cannot also be a proficient pilot, as long he realizes that proficiency doesn't happen automatically and he is willing to work at acquiring and maintaining it.

Flying is not a "normal" thing for human beings to do. In fact, it is a potentially very dangerous thing to do. What makes it safe is pilot proficiency, and proficiency is not something you are born with, although your ego will try to tell you otherwise.

Airplanes are supposedly dumb, mechanical beasts, but they have an uncanny way of finding and punishing pilots who let their egos control their actions.

PROFICIENCY

Airplanes also have a way of rewarding pilots who respect an airplane's capabilities and dangers. That reward comes in the form of fast, reliable transportation which in turn leads to enormous self-satisfaction. And self-satisfaction is the best ego-trip of all.

11
Judgment

I DIDN'T SLEEP MUCH THE NIGHT BEFORE I TOOK THE TEST FOR MY PRIVATE PILOT'S license. I was nervous to begin with, and the nearest examiner was 60 miles away, which meant that before I could even take the test I had to fly cross-country to an unfamiliar airport, so I had a lot to think about.

I wasn't up to confessing my anxiety to my flight instructor, but because failure seemed to be the most likely outcome, I thought a little bit of prep work was in order. So when I got to the airport I said something stupid like, "I sure hope he isn't big on turns-about-a-point or any of that kind of stuff."

I was looking for reassurance of course. I should have kept my mouth shut. If my instructor had said, "I wouldn't count on it. He loves to do stuff like that. You might as well just throw up right now and get it over with," it would have been bad enough. What he said was worse.

"Don't worry about the airwork. He assumes you can do the maneuvers. What he wants to test is your judgment."

"My judgment? I think I left that at home."

What's "judgment?" I have two plotters, three computers and four pencils, and he wants judgment. I had never even heard the word used before in an aviation context. All the time I had spent to get to this point, the great private pilot flight test, and I had never once heard the word "judgment" used and it turns out that's what he wants to test. This is going to be funny, I thought. I can't wait.

Somehow I got the little airplane over to the appropriate airport and the flight test turned out to be very typical and predictable. We did the usual assortment of airwork, cross-country procedures, and takeoff and landings. I did some of it well, some of it not so

JUDGMENT

well, some in between, and the examiner, whose name was Santa Claus if I remember correctly, signed me off.

But that word bugged me for years. *Judgment*. It had the aura of something vague and mystical, maybe even something only rare individuals were born with—something the rest of us mortals could only envy, like "charisma" or "brilliance." I knew it had something to do with maturity, and that was good, because I could only get more mature, but would that be enough? Somehow I knew it wouldn't be—somehow I knew that "judgment" was more than just maturity.

INSIGHT

Finally, one day, years later, sitting in my car waiting to pull out into traffic, it hit me what "judgment" was. "Judgment" was knowing whether to shoot out in front of the next car or wait for a bigger opening. It was as simple as that. After years of driving, it was something I did unconsciously, something we all do unconsciously. Every time we pull out into traffic we unconsciously make a decision, a decision based on our experience, our skill, our training, and our discretion. We call the sum of that decision-making process our "judgment." Terrific. I'm ready to take the test now, sir.

This was a big breakthrough because it took much of the vagueness out of judgment. It meant that judgment wasn't something like wisdom, that maybe, if you were smart enough, and you lived long enough, you might hope to obtain but probably wouldn't, and couldn't do anything about anyway. Judgment was something that, to a certain extent at least, could be learned and acquired. In fact, the more training you took and the more experience you got, the better your judgment should get (in theory, at least).

JUDGMENT AS LEARNING

A brand-new student driver has very little idea when to pull out into the traffic. He hasn't learned yet to estimate distances and speeds—closing rates—nor does he have a good "feel" for how the car responds: "If I step on the gas, can I count on the car to go right now, or is there a hesitation or sluggishness to contend with?" The student driver has neither good nor bad judgment—he has *no* judgment.

He is dependent upon the instructor telling him when it is okay to go. After some instruction and some practice, he starts to get a better idea for himself—he starts to acquire some judgment. When he goes to take his driving test, this is one thing the examiner looks for: does he have a good sense of when its safe to pull out and when it isn't?

It is a skill in the sense that it is something that can be learned and something that gets better with practice and experience, but it is called "judgment" because it isn't just a physical skill. It also is a mental process. A *decision* has to be made.

FLYING JUDGMENT

Flying has many analogous situations. In fact, flying is not so much a series of physical manipulations as it is a series of judgments. The physical part is usually the easy part, but that's the part in pilot training that gets all the attention. How hard is it, physically, to "go around," for instance? Not hard at all. All you have to do is push the power lever all the way up and change the pitch attitude, then raise the gear and flaps. But making the decision to go around—exercising that judgment—is the hard part.

Exercising good judgment is what being a good pilot is all about. But that's not what most people think. Most people think that the physical manipulation of the controls is what being a good pilot is all about. That's why the passengers always pay so much attention to the landing. They think that the landing is the most difficult and critical part of the flight, and if the pilot does that part well he must also do everything else well. It just isn't so. The most difficult part of the flight is not the landing; the most difficult part of the flight is making the proper judgments to arrive at a point where a safe landing can be made.

Certainly there are physical demands made upon pilots: gusty crosswind landings, circling approaches to minimums, engine failures, night approaches over water. All these situations require considerable physical skill. But by and large, once you have mastered the basics and have enough experience to be comfortable with the airplane, the physical act of flying the airplane is the easy part.

Professional pilots call the guys who are physically skilled at manipulating the airplane "good stick-floppers." As you can probably tell from the choice of words, it is not an unequivocal form of praise. This isn't because pilots don't respect guys who are naturally skilled at flying the airplane, it is because the description "good stick-flopper" is usually used to "damn with faint praise."

It frequently means something else more important is missing, like experience, maturity, or judgment. If a pilot is truly a good pilot in all respects, another pilot will simply say, "He's a good pilot." Normally, "He's a good stick-flopper" translates into: "He handles the airplane well, but his decision-making ability leaves something to be desired." It is the aeronautical equivalent of describing a blind date as having a good personality.

I toss that out about the "stick-floppers" to illustrate the importance professional pilots place on judgment verses physical skill with the airplane. Because an airplane is an unwieldy and unforgiving sort of beast—much more so than an automobile, for instance—the physical aspects of flying get a lot of attention, and rightly so, at least in training. It isn't easy to learn to fly and there are always things you can learn to do better, and areas that get rusty if you don't stay in practice. But in terms of the priorities for safe flying, judgment—the part you do with your head—ties up approximately the first 90 percent, and flying the airplane—the part you do with your hands and feet—gets the remaining 10 percent.

GRAY AREAS

The part about judgment that is so hard to get a grip on—and this is what threw me for a loop about it for so long—is that decision-making in aviation is not black-and-white. It seldom is in any endeavor of course, but some things are much more black-and-white than others. Deciding whether to buy or rent an airplane, for instance, is a fairly straightforward process of projecting usage and playing with some numbers.

JUDGMENT

The final decision (assuming emotional considerations can be left aside), doesn't involve much judgment. But because judgment in aviation is seldom black-and-white, it can't be reduced to a mechanical process, or handed over to a computer, or simplified into "rules-of-thumb." Only a human being can exercise judgment.

Judgment is a very complicated thing. Judgment can be learned and experience helps a lot, but ultimately a human being has to make a decision that in many cases could be argued either way—there might not be an absolute right or wrong to it. Even the final outcome won't necessarily prove the decision right or wrong. Lots of things turn out all right, even though the decision behind them was wrong. The kid, for instance, who pulls out directly in front of a loaded cement truck, tires squealing and rocks flying, and just makes it, still made a lousy decision. He exercised poor judgment, but because it turned out all right, it is hard to prove that what he did was wrong. About the only time a decision is *proven* wrong is when things actually *do* turn out badly—the cement truck crunches him—and, fortunately, that is rare. Sooner or later bad judgment tends to catch up with you, which is why teenage drivers have such a terrible driving record compared to older drivers, but the fact that nothing has gone wrong yet doesn't prove the existence of good judgment.

ESSENTIAL ELEMENTS

Perhaps the best way to get a grip on what is, after all, a fairly slippery question, is to look at a couple of flying situations requiring the exercise of judgment, and examine the part each component—knowledge, experience, skill, and discretion—plays in that process.

One area where judgment plays a major role is avoiding wake turbulence. One of the first things a student pilot learns is that all airplanes generate wing-tip vortices that are very turbulent and the degree of that turbulence is in direct proportion to the weight of the aircraft, and is in indirect proportion to the speed. Therefore large airplanes, on final, heavy and slow, will generate the most severe wake turbulence, and are to be treated with the most respect. But the exact degree of "respect" is a matter not of law nor science, but of judgment.

The controller's rule is to allow at least three miles between any two airplanes, increasing the separation to five miles if the airplane in front is a "heavy" (more than 300,000 pounds). But these are mere minimums for controllers to use. As the pilot, in visual contact with the traffic ahead, you can increase or decrease this separation at your discretion. How do you decide?

Several decisional elements are involved. There is the technical element: knowledge of wake turbulence theory. There is the element of experience: personal, first-hand knowledge of wake turbulence. There is the element of pilot skill. And finally, there is the intangible element of discretion. Whenever "judgment" is involved, there comes a point after the knowledge, experience and skill have been plugged into the equation, where a human being has to exercise his discretion, and the decision that results reflects to a large extent what kind of person he is.

Essential Elements

The first step is the knowledge part. This cannot be skipped, but often is. Often a pilot goes straight to the question of how much distance he feels is appropriate between himself and the "heavy" in front based on a purely arbitrary decision or guesswork or what he just read in a magazine or heard somebody say. If he doesn't feel any bumps he says he exercised good judgment, and if he does feel bumps, he says the guy in front slowed up and didn't tell anybody.

You can't make judgments arbitrarily—not good ones anyway. Good judgment has to be based on fact to make any sense, so the first step, in this case, is to find out as much as possible about wake turbulence and what the best current thinking is to avoid it. This means reading all the available literature including the AIM and the appropriate sections from a flight training handbook, and it means having received instruction in the techniques of wake turbulence avoidance. This then forms a basis of fact and knowledge about the subject so your decisions can be based on reality and not myth, or wishful thinking, or the war stories of the heroes who hang around the airport lounge.

The next part is experience. With a knowledge of the facts and theories on the subject under your belt, you have the basis for forming sound guidelines for avoiding wake turbulence: three to five miles separation and always above and upwind of the glide path of the aircraft in front. Experience will then reveal whether those initial guidelines are good ones or not.

Experience is merely self-instruction—the "School of Life." One smooth, quiet morning, when you get rolled 90 degrees following a DC-9 by seven miles, you will know that even five miles doesn't always do it. Another day, when it is not so smooth and you are following a 747, and you accidentally drop below the glideslope and don't feel any wake turbulence at all, you will have learned that turbulence can do a pretty good job of breaking up wing-tip vortices. (Or maybe you just got lucky and both you and the 747 dropped under the glideslope at the same point.) This doesn't mean you can disregard considerations of wake turbulence when it's bumpy, but it does mean you have learned, from personal experience, how to modify the guidelines—allow extra room when it is smooth, and turbulence can, in some cases, be beneficial. Experience has improved your judgment.

Pilot skill is the least important factor in exercising good judgment. If you are flying an airplane that you have hundreds of hours in and are very comfortable with, and you have experience with wake turbulence in that airplane—maybe touching the burbles of turbulence around the edges once or twice—then maybe you can safely accept the minimum separation. But no amount of skill is going to make you immune to the dangers of wake turbulence. Pilot skill is the least important variable in the judgment equation.

A known lack of skill, on the other hand, can cause you to modify personal judgment substantially. If you've only flown a couple of ILSs before, and the controller has you exactly three miles behind a jet and you aren't completely confident you won't drift below the glideslope slightly, you could tell the controller that you would like more separation. He can do it as long as you don't wait until the last minute to pop it on him. It might cost you an extra turn or two, or he might even have to take you out of line and bring you back

JUDGMENT

in again at a point where there are no jets in front of you. In the worst case it might cost 10 minutes. If you know you lack skill in a certain area, then good judgment dictates extra caution. Ten minutes is a very small price to pay for a smooth and safe flight.

The part that makes judgment different from simple decision-making is the human element. Whether we call this element maturity, discretion, or wisdom doesn't really matter. The fact is, because exercising judgment is not a matter of black-and-white or right-or-wrong, the judgments that are made reflect, to a large extent, the personal makeup of the pilot involved. To the extent a name for that personal element is necessary, call it "discretion."

That all goes back to the oldest cliché in aviation: "There are old pilots and bold pilots, but there are no old and bold pilots." Some pilots are bolder than others. Two pilots can have identical training, experience, and skill, and one will decide that three miles behind a "heavy" is okay but the other wants five miles. As long as both land without incident, both decisions are correct. But the one five miles behind showed better judgment. It is, of course, possible to be unnecessarily conservative. Ten miles behind would probably be excessive, but no one can argue with five miles. The guy who consistently elects the "bolder" course of action is consistently exposing himself and his passengers to a higher degree of risk, and in so doing is reducing his chances of making it to old pilot status.

This human element is the variable that makes judgment such an unwieldy concept. I can't change your personality, and regardless of how mature you are, you are less mature at your present age than you will be 10 years from now and there is nothing you can do to accelerate that process. All I can do is identify for you the elements that make up judgment, encourage you to do the best possible with the elements that are under your control—knowledge, experience, and skill—and hope that you either know already or will someday learn the virtue and wisdom of following the path that is less bold. In this specific example, that means following three miles behind any jet and five miles behind a "heavy" is an absolute minimum and any separation you can get beyond that is better and, in some cases, essential.

Let's look at another example. You have just completed an ILS approach and have broken out at circling minimums. The tower has provided the option of landing straight-in, but with a gusty, 20- to 30-knot crosswind, or circling to a shorter runway that is aligned with the wind. The airport is surrounded by mountainous terrain and the circle will be tricky, with minimum visibility (two miles), and the approach end of the desired runway is hidden by a hill for the first half of the circling approach.

This is a tough one, because neither choice is very good. (Unfortunately, that's the way it usually is in the real world.) Many elements are at work here. Nobody likes to circle unless absolutely necessary because it takes a lot of concentration and work to do a good circling approach under the best circumstances and this is the worst circumstance.

Flying along just under the clouds and just above the terrain at a relatively slow airspeed is inherently riskier than a straight-in approach. It's very tempting to take the easy way out and just go straight in. But in so doing you take on a gusty crosswind that is difficult and somewhat risky. So what is the "right" thing to do? What does the "Smart Set" do under these circumstances?

Let's break it down into the essential elements.

Knowledge. You know that a circling approach, if flown properly, provides marginal but still safe separation from the terrain. You know that the degree to which it is safe or not safe is completely dependent upon your skill as a pilot. You know that a mistake could well be disastrous. You also know that gusty, crosswind landings are tricky and might result in damaged aircraft ("pranged" is the euphemism most commonly encountered).

Experience. You have done a few circling approaches to minimums in the past. They have all turned out well, but experience says that they are demanding and unforgiving, and that they are particularly disconcerting when the approach end of the desired runway cannot be seen until near the end of the circling maneuver. You have a lot of experience with crosswind landings and know that there is a big difference between a steady crosswind and a gusty crosswind, and that for a gusty crosswind, the size of the gust factor is the key determinate. Experience says that 20 knots gusting to 30 is right at the limits of what you and the airplane can handle.

Skill. You feel equally skilled in each maneuver, but a higher premium is placed on skill in the circling maneuver. Because you don't feel especially skilled in one over the other and are confident you could, if necessary, do either safely, skill is not a factor in this case in exercising your judgment, but you lean towards the crosswind landing.

Personal Element. What kind of person are you? Are you conservative and careful and don't care much if the passengers think you're a lousy pilot because the crosswind landing you attempt is bumpy, landing the airplane "crooked" (on the upwind wheel) and setting down hard? Or are you a pilot who takes real pride in a successful circling approach under difficult conditions, especially if you can give the passengers a smooth ride with a "squeaker" landing? Do you worry about the consequences if worst comes to worst and decide that a damaged airplane is better than a disaster, or do you think that question is irrelevant because you are completely confident that you can handle either situation? Are you the impatient type who wants to get on the ground as fast as possible, or do you enjoy the challenge of a tough circling approach?

Let me say quickly that this situation can be argued both ways. If you are very knowledgeable, experienced, and skilled at circling approaches, good judgment might dictate that you elect the circling approach, thus eliminating the difficulties of the crosswind landing. (The problem with gusty, crosswind landings is that even the most experienced pilot can be "embarrassed" once in awhile because no one can "see" the wind and the right one-two punch of gust and sheer can catch anybody.) But for most pilots the crosswind landing would be preferable because the "downside risk"— the worst case scenario—for the crosswind landing is so much less than it is for the circling approach. And just as there are good reasons for doing it either way, bad judgment is anything you decide on the basis of laziness, wishful thinking, a desire to impress your passengers, or impatience.

What if the crosswind is 30 gusting to 40? Now you *must* circle, right? Perhaps not. You could fly to the alternate, or, if the condition is temporary, as it might be if a thunder-

JUDGMENT

storm were passing near the airport, you could go somewhere else and hold awhile. "Good judgment" sometimes means remembering there are other choices.

THE HARD CALLS

The two toughest situations for making judgment calls are: when nobody else is around to help you out, and when everybody else is doing something else. Let's look at two examples.

You show up at the airport and the runway is covered with three inches of light snow. All the plows are broken and the airport manager says it will be at least three hours before he can start plowing. You have never taken off on anything but a plowed runway before and have no idea what effect the snow will have on acceleration and aren't too anxious to try to find out. The local 19-year-old flight instructor says "Hey, give it a try. The snow's dry . . . it won't bother you. You shouldn't have any trouble at all." The chief mechanic, who doesn't fly but who has formed an opinion or two in his 30 years of turning screws, just shakes his head and walks off.

Nobody has taken off since the snow started to fall. You have nothing to go by. One person thinks you'll have no problem and another thinks you're crazy. If you don't go, one of them will think you're a sissy; if you do go the other will think you're a fool. Wouldn't it help if somebody else took off first? Then if it works out without any problem, you could follow him and act like you planned on it all along, and if he ends up in the snowbank at the end of the runway, you can thank your lucky stars you were born with the maturity and good sense to know not to do something as stupid as that.

So what do you do with a tough call when you have no experience to guide you and no example to follow? The answer is obvious: take the safest way out regardless of the cost or inconvenience. The advice of others who are not in your place is cheap and unreliable. Putting up with the snickers of the resident experts is no fun, but you have to play your own game.

The same principle applies when everybody else is doing something you have misgivings about. It's awfully hard, for instance, to cancel or postpone a flight into possible icing conditions when everyone else is going, but if it doesn't seem right, don't do it. I know I sound like your mother telling you that "Just because everybody else does it doesn't make it right," but every now and then a guy makes a true hero out of himself when he does not follow the pack.

I have made the opposite of a true hero out of myself, on occasion, when I have not followed my instincts, and it's not a nice feeling. One night at LaGuardia, after waiting in line for takeoff for close to two hours, I took off into a whole bunch of thunderstorms. It didn't seem right at the time, but plane after plane had gone in regular procession in front of me, and I hadn't heard a single adverse report, so I figured that the situation in the air had to be better than it looked on the ground.

The takeoff was fine—fairly smooth—we didn't even start to lose ground contact for 4,000 or 5,000 feet. The corporate jet's radar showed an area of heavy precip directly over the departure path, but everybody ahead was going through it with no problem, so again I rationalized that it had to be all right. It wasn't. The only warning was two small bumps, and then, for about 15 seconds, the kind of turbulence that, even with the belts tight, lifts you right out of the seat and throws everything all over the cabin.

We were tossed around so much that I lost my grip on the control column, couldn't get a hold of it again, couldn't get to the autopilot button, couldn't see the radar (not that I could have done anything about it even if I could have seen it), and got hit first by a mike and then with a Jepp book. After that it was smooth sailing for the rest of the trip, a good part of which was spent trying to explain what happened to our only passenger (who just happened to be the corporate officer directly responsible for the flight department).

All I could do was apologize for the awful ride and tell him that no one else had had any problem, but he knew and I knew that the explanation was weak. Between the time the airplane right in front of us went through that area and the time we did—a matter of three or four minutes—the cell must have built up to severe proportions. I would like to have been able to say that it was just "one of those things"—an act of God—that was unavoidable. But it wasn't.

If you takeoff into thunderstorms, you have to expect that once in awhile some "funny" things will happen. The fact that "everyone else" seems to be cheating the odds can't change that. I exercised poor judgment that night (and so did a bunch of other pilots who happened to get away with it, but that's not the point), and I got caught. It's tough to say "no" when everybody else is doing it. Watch out for this one. It's a real sucker hole.

CONSERVATISM

When it comes to making good judgments, knowledge and experience are invaluable. No matter the situation, the pilot who knows something about the situation and has some first-hand experience with it will be in a position to make a better decision than the pilot who doesn't. In addition, the more experienced and knowledgeable the pilot, the more likely he is to make a conservative decision.

Experience almost always tempers judgment in a conservative direction, and knowledge almost always leads to an awareness of hidden dangers, which in itself leads to conservatism. Because there are no short-cuts to experience, the only way to accelerate the process of acquiring good judgment is to expand your knowledge and to be conservative, even if you don't fully understand why. Experience will, in time, teach why being conservative was smart.

When I first started flying professionally, I frequently found myself questioning the decisions of the captains. I've never been a "hotdogger," but sometimes it seemed like the "old guys" deliberately went out of their way to be conservative—unnecessarily so. After all, enough is enough. But the more I flew, the more I found myself agreeing with these guys. They were still fairly conservative, but they didn't seem to be *unnecessarily* conservative anymore.

Now that I have had the chance to be a captain awhile, I find that it's the quality of some of the copilots that seems to be the problem. Some of them are still too anxious to take chances when they don't know what they're doing, and I can tell that they think that

JUDGMENT

I'm being too conservative when I overrule them. But I do anyway. Being conservative is a captain's privilege. As the pilot-in-command, it is your privilege also.

FLIGHT TESTS

Examiners do try to test judgment, but testing judgment is a very hard thing to do. An examiner can test your knowledge, and your skill, and can assess your experience, and in so doing can get a rough idea of what your judgment is likely to be. It is very hard to find out what kind of person you are in the space of a short flight test, and without that information, an accurate assessment of your judgment is impossible. Despite what my flight instructor said, examiners don't test judgment. Airplanes do. Good luck on your test.

12
Strategy

Few pilots can say they look forward to a flight physical, but when the doctor giving the physical is Dr. Dick Cardozo, that is usually the case. Dr. Cardozo is now retired as the head of the Hitchcock Clinic in Hanover, New Hampshire, and no longer does FAA physicals (giving him that much more time to fly his pressurized Cessna Skymaster around with his partner and able successor, Dr. Nield Mercer), but when he was performing my physicals, I always looked forward to it because I never failed to learn something from him—he once took 15 minutes to try to explain EKGs, for instance. Usually the things I learned from him were not directly medically related, though. Usually what I learned from Dr. Cardozo was matters that simply came up during the course of conversation while being banged on the knee, poked in the stomach, and so on.

I was told that Dr. Cardozo's father (or was it grandfather?) was the famous Supreme Court Justice Cardozo, and that Dr. Cardozo performed the first open heart surgery in New England. I probably should try to verify these facts with him, but I won't because I know he won't give me a straight answer because he is too modest and I prefer to simply believe they are true and have no reason to think they are not. The very first time I saw Dr. Cardozo was in the late '60s. I remember it well because he was climbing out of a classic, 12-cylinder Ferrari. He sold it when he realized that it made no sense to drive what was essentially a race car on anything other than a race track. (Still, I'd love to have that car.)

Dr. Cardozo is a vegetarian, a student of Hindu mysticism, and he has hiked extensively in the Himalayas. Still, the classic manner of the traditional family physician comes to him naturally, greeting every patient—new and old alike—with a smile, an extended hand, and a big, "Hello, friend." He also happens to be a very capable and conscientious pilot who takes his flying as seriously as his medicine.

STRATEGY

During one of my first physicals with Dr. Cardozo, he asked me what I did for exercise. I sort of expected this question and I was feeling a little defensive about it because I have always been one of those lucky people who has to eat four or five times a day just to get up to skinny, and the fact was I didn't really do anything about exercise. But I didn't think I could just come right out and say that, and I wanted to do my best to avoid the lecture that I thought must be coming: "Well, I have a 10-speed bike that I ride some. I should ride it more, I know, but I do ride it some," which was almost true—I probably rode it at least once a month. I then sat back and and waited for the lecture. He looked at me and said, " 'Should' is a super-ego word." That was a good one. First of all, I didn't know what "super-ego" meant, but it sounded like it was either real good or real bad. Second, if I didn't know what "super-ego" meant, I sure didn't know what a super-ego word was. But I didn't ask, because if that was as bad as the lecture was going to be, I thought I could live with it.

I did look up "super-ego" at home, and I learned that the super-ego, in classical Freudian theory, has to do with those things our parents, teachers, and society in general tell us we ought to do, things that we, as the normal, ornery rascals most of us naturally are, wouldn't do unless someone made us. (The super-ego is supposed to balance the id, which is all that normal, ornery stuff that makes us rascals.) When your mother wouldn't let you go swimming with all the other kids and made you go visit your great-aunt instead, and when it was finally time to go you had to say you enjoyed it and let the aunt kiss you, that was super-ego training.

I told Dr. Cardozo that I knew I "should" ride my bicycle more, and he told me that "should" was a super-ego word, which I now saw was obvious—almost a tautology. (Look it up—I can't do it all for you.) "Should" is how all that coercive and conforming information—all that parental stuff—gets transferred, which meant that he wasn't lecturing me at all. In fact, he had done the exact opposite, saying, in effect, "I'm not interested in what you think you should do, or even what you think you're supposed to think you should do. I just asked you a simple question. A simple answer would be, "Nothing."

I think about that a lot—about all the times people tell other people what they *should* do and how after awhile we end up repeating all the "shoulds" almost automatically ourselves whether we really believe them or not. I have come to the conclusion that on the day we are born it might be a good idea to issue personal books with 10 "should" coupons inside. Everyone would have to carry this book around at all times, and each time the word "should" was used, you would have to fork over one coupon. When all the coupons were gone, you'd be through and your "super-egoing" days would be over.

The problem for me, though, is that it is very hard to write a book that I hope will help people with their flying without using all 10 coupons in about the first three pages. How do I tell you what I think is important in flying without telling you what I think you should do?

I don't know. It puts me in a spot. So I've settled on a deal of sorts—a typical, American compromise. I will tell you about a five-part strategy for safe flying that I have developed. Five key steps that I am quite certain will virtually assure you safe and uneventful flights, but I won't demand observance. I won't even say you *should* observe them, I will just leave that to your judgment, maturity, and good sense.

But if you decide they do not make sense for you, then I *will* tell you that you should observe them, which will still leave me with five "shoulds" in my book, enough, I hope, to get me through a few more good years anyway.

Actually, to be a little more serious (I remain hesitant to tell people what to do, but I don't know how to write a book on safe flying otherwise), I got the idea for forming a five-part strategy for safe flying from Arnold Palmer. I'd love to say that I got the idea from him while banging the ball around Latrobe after he'd had me over to check out his flight operation, but the fact is that I got it from a book he wrote on golf, a book called *Play Great Golf.* I love to play golf, and I have a bunch of books on the subject, and while I have enjoyed reading all of them, none of them has ever helped my golf game very much except for Palmer's book. The reason for that is, I think, that Arnold Palmer doesn't try to put down every single thing he can think of regarding how to be a good golfer, he simply lists and describes five very specific things all good golfers do regardless of their individual swing style, five things that Palmer believes anyone must master if he or she wants to be a good golfer.

The reason I liked this approach was that I knew I couldn't take a whole book of ideas to the golf course, and even if I could, trying to swing like Greg Norman says to swing in his book might never be right for me because I am not built like Greg Norman and I do not have his natural athletic ability.

I can take five ideas to the golf course—five things that are common to all good golf swings—and I can work on (I can even remember) those five things:

1. A good grip.
2. The proper address.
3. A smooth start to the backswing.
4. A steady head.
5. Acceleration of the club through the ball.

So I thought, Why can't the same thing be done for aviation? Wouldn't it be terrific if "great flying" could also be boiled down to five specific, easy-to-remember items? I'm not absolutely sure it can, but I have tried to do that anyway, and the result is the five-part strategy that follows, a strategy called the BILAHs (bylaws) for safe, cross-country flying.

BILAHS

As you might well guess, the capital letters in BILAHs represent the key words for each part of the strategy. The "s" on the end comes from the last key word, and is just there to make the word "BILAHs" easier to say and remember. Pilots love memory aids—GUMP for gas, undercarriage, mixture, and propellers is the pre-landing check and "East

STRATEGY

is least and west is best," the correction for magnetic variation, are probably the two best-known aids. There is a lot to remember in aviation and memory aids do help, so I have tried to do my part.

I certainly don't expect BILAHs to become as famous as GUMP, but I did think something was necessary to help in retaining the strategy once this book has been put away.

(If you have read *Fly Like a Pro,* then you know that I made a stab at this same idea, which was before I read the Palmer book, but the idea had not clearly formed at that time and I tied it into the weather briefing, rather than an overall flying strategy, so it wasn't quite the same thing. But the weakest part was that the best I could come up with for a memory aid was IBFAHs, which was probably harder to remember than the five items themselves. I hope this is an improvement.)

The five words represented by BILAHs are:

B: Briefing.
I: IFR.
L: Log.
A: Alternate.
H: Hazardous weather.

Observing them is essential to good, safe, cross-country flying.

BRIEFING

The "B" in BILAHs is for briefing, weather briefing. I couldn't label the first step "weather briefing" or the memory aid would then be WILAHs—this isn't as easy at it looks. Ask for and receive a complete standard weather briefing prior to every cross-country flight, regardless of how short or familiar the trip is, and regardless of how certain you are on the basis of other information that the weather will be all right.

Recall that it is very hard to make bad decisions about the weather with good information—not impossible, but hard—but it is very easy to make bad decisions with bad or incomplete information. In any case, it is virtually impossible to do a good job of flight planning without a good weather briefing. It costs nothing except a little time to get a good weather briefing, and even if all it does is confirm what you knew or thought already, that is fine and there is no harm done.

There is no good reason whatsoever for not getting a complete weather briefing before every cross-country flight, but many bad reasons for not doing so: laziness, overconfidence, ignorance, arrogance, being in a hurry. A good weather briefing is absolutely fundamental to safe flying for all pilots and for all operations, and has to be one of the five keys in any strategy meant to ensure safe flying.

IFR

The "I" in BILAHs is for IFR. The single best thing any pilot can do to ensure the safe and uneventful outcome of his flight is to file IFR. I get more disagreement in meetings and conversations with general aviation pilots, both instrument-rated and non-rated (more disagreement from rated pilots, in fact, than from non-rated, who don't always know yet whether they agree with me or not), on this one item than any other. I have listened to all counterarguments, they have listened to mine, and very few minds have been changed, but I still think it is true.

My argument for filing IFR at all times, regardless of whether or not the weather requires it, is based on the belief that the many, many advantages of filing IFR far outweigh any disadvantages, and that the only way to be comfortable with the instrument flying system is to file IFR every time. If you're a little hesitant to file IFR when the weather is good, how comfortable are you going to be filing IFR when the weather is bad?

What are some of the advantages of IFR over VFR? Good question.

Here's a list of 10.

1. Freedom from the need to avoid IMC en route (inadvertent entry into IMC by non-rated pilots is one of the most frequent causes of accidents among general aviation pilots.)

2. Virtually continuous radar contact. What are the advantages of radar contact? Another good question. Here is an additional list of five:
   a. Flight following.
   b. Traffic advisories.
   c. Automatic separation from restricted, prohibited, and warning areas.
   d. Severe weather advisories.
   e. Navigational assistance.

3. Automatic entry and exit from TCAs and TRSAs.

4. Continuously available emergency assistance.

5. Standardized flight planning using preferred routes, standard instrument departures and arrivals, and detailed en route charts.

6. Automatic terrain clearance and avoidance.

7. Simplified customs and ADIZ reporting.

8. Automatic flight plan opening and closing.

9. Easy transitions from the en route to the approach- and tower-controlled phases of flight.

10. Easily locating destination airports.

STRATEGY

What are the disadvantages? Lousy question, but I'll answer it anyway.

1. Restrictions in altitude and route selection.
2. Routings that are less direct.
3. Diminished freedom to change altitudes and headings as necessary to avoid areas of adverse or hazardous weather.
4. Pilot must be instrument-rated and current, and aircraft must be appropriately equipped for proposed flight.
5. Required flight plan filing 30 minutes prior to departure.

The advantages clearly outweigh the disadvantages, but what I get from skeptical pilots is an argument:

> Well, sure, a lot of the time it makes sense to file IFR, but not all the time and it sure can be a pain—all those crazy clearances with vectors and altitude changes and routings that take you all over the countryside except where you want to go and then you have to hold . . . and besides, the last time I tried to go IFR the controller tried to get me to go right through a whole line of cells.

My counterargument is that it's only a pain because you don't do it all the time. If you were to file IFR every single time you fly outside the local area, you would very quickly get used to filing instrument flight plans and having to make an occasional altitude or heading change (and almost no one has to hold anymore except around the busiest airports, nonetheless probably nothing is dreaded more by the inexperienced instrument pilot than holding), and you would very quickly begin to see that there are always reasons for those changes; reasons mainly related to traffic avoidance but sometimes for other reasons like avoidance of restricted areas, and why wouldn't you want to avoid traffic or a restricted area anyway? The controller is actually doing part of the work for you when he issues those changes.

It's the last argument, though, where they think they really have me:

> How do you stay out of weather you don't want to go into if you're IFR and the controller tells you to go there, and how can you argue that IFR is better when there are thunderstorms and buildups all around, you can see them and he can't, and it would be a relatively simple matter to stay out of them VFR and a very difficult matter IFR? Gotcha there don't I, Mr. Expert. Seems like you must have come down with a sudden case of lockjaw. You don't have a good answer for that one, do you?

Well, of course I have a good answer. First, what are you doing out there with buildups and thunderstorms all around? Is poking your way through black, boiling clouds your idea of a good way to get from A to B? Most examples pilots have given me of how they flew IFR into a bunch of trouble with rain and thunderstorms and had been vectored or cleared into rotten weather are good examples of flights that shouldn't have occurred in the first place (not with unpressurized, no deice, and non-radar equipped aircraft at any rate), or, as a minimum, should not have occurred anywhere near that route of flight.

Second, even if you do confront a bunch of nasty weather, no pilot has to accept a clearance. If you don't want to fly heading 080 directly into a cell, don't do it. Reply, "Unable heading 080. Weather." You might have to wait while the controller figures out Plan B, or go some other direction that isn't where you want to go but doesn't have adverse weather. You do not have to go where cleared unless you have read back the clearance, indicating that you have received and have *accepted* that clearance. Until you do that, you don't have to do it.

So I don't buy it. I still think IFR is the only way to go, I don't think it is a pain, not much of a pain if you get used to it by doing it all the time. Any pain is a small price to pay for the many advantages.

There might well be times when you shouldn't go at all, VFR or IFR, and there might be times when you have to plan a more circuitous route, but I have yet to hear a convincing argument for a situation that was safe to fly VFR when it was not safe to fly IFR.

I say again, the single best thing any pilot can do to ensure safe and routine cross-country flying is to file IFR on every trip. If you aren't instrument rated yet, I recommend that you start on that rating right now. The benefits will be immediately available in the form of increased confidence in your flying and increased capability to handle more-complex situations, and, in the meantime, you can fly VFR as if you were IFR simply by flight planning along airways and by asking for radar advisories from takeoff to touchdown, and in so doing reap many (but not all) of the advantages of IFR.

LOGS

The "L" in BILAHs stands for "logs," and that means "flight logs." I'm a big believer in flight logs and between what I said regarding flight planning and flight logs, I have argued the case for flight logs so many times and in so many different ways that I am in serious danger of repeating myself one time too many.

The simple fact is that a good flight log virtually assures good flight planning, and good flight planning goes a very long ways toward assuring safe and uneventful flights. Having a good flight log is also the best way to keep tabs on a flight en route because the information in the log provides a benchmark against which the actual flight data can be compared and positive or negative trends observed.

Finally, flight logs are the best way to get to know an airplane because the preparation forces you to make certain assumptions about aircraft performance (airspeed and fuel flow) and the maintenance of that log in-flight confirms, in irrefutable detail, the validity, or lack of validity, of those assumptions, and that's really what "getting to know an airplane" means: how does it actually perform? So, for many very good reasons, flight logs are very important to safe and routine cross-country flying.

STRATEGY

ALTERNATES

Having an alternate for every single flight, VFR or IFR, good weather or bad, is the single best way to be sure you never run out of gas. In fact, it is the single best way I know to ensure never even having to worry about running out of gas. (Much better than simply filling the tanks with fuel.)

"Having an alternate" means identifying a secondary destination with weather forecast to be well above approach and landing minimums at or about the time you expect to arrive, if that becomes necessary, and (it goes without saying but has to be said at least once anyway) ensuring that you have the fuel to reach that alternate. In simple terms, "having an alternate" means having a bailout option—a place to go—for those times when the intended destination is closed due to weather or operational restrictions such as runway closures, fires, or other emergencies on the field.

For flights under VFR, there is no requirement to have an alternate because the FAA assumes the existence of an unlimited number of alternates whenever the original conditions are suitable for VFR. This is usually a fairly valid assumption; if the route over which you plan to operate is VFR, and if your intended destination is VFR, then presumably any number of airports along that route and in the general vicinity of the destination will also be VFR.

It is still a good idea for the VFR-only pilot to designate and fuel for a specific alternate airport, one that is forecasting good VFR, if for no other reason than to ensure that he has at least thought about the possibility of not landing at his intended destination and the consequences thereof. That does happen, and thinking about and planning for it ahead of time goes a long ways toward ensuring that it does not become a critical event.

Mainly, when I say always having an alternate is one of the five keys to safe flight, I am referring to operations under IFR. By regulation, you do not always have to have an alternate when IFR, but I think is important to have one anyway, regardless of the weather. One of the best reasons for this is simply convenience: the regulations that describe when an alternate is required and when it is not (destination forecast must be at least 2,000 and three for two hours before and after the ETA) are hard to remember, easy to get wrong, and, in any case, are not quite as conservative as I would like for something as important as this. I am not comfortable taking off only with enough fuel to get to my destination plus 45 minutes when the destination forecast is no better than 2,000 and three. So, for several reasons, it makes sense to always have an alternate for every flight, regardless of the forecast weather. (Most countries require at least a single alternate and some countries even require two for all IFR operations. The FAA regulations are very liberal in this regard.)

In fact, it makes so much sense that this is the part of the strategy almost everyone agrees with because it seems so simple and everybody likes to take lots of fuel anyway and this helps provide ample justification for doing that. That is, everyone agrees with me until the time comes when they have a long trip with a destination forecast to be better than 2,000 and three, but they can't find a suitable alternate within range because virtually all the fuel capacity is needed just to get to the destination and still have 45 minutes of fuel left in the tanks.

They find themselves in the position of being able to go but only if they are willing to waive the self-imposed "always have an alternate" rule. Suddenly it doesn't seem like such a great idea anymore, and so they go, sweating it out the whole way, hoping and praying that the destination weather doesn't go down.

Of course, if they are regularly checking the destination weather en route and it starts to drop early, they can always make a precautionary landing, but that's not what I call a safe and uneventful flight. Who wants to spend a whole flight checking the weather every chance you get, watching the fuel gauge more than the VOR needle, and keeping tabs on all the available airports as you fly by them?

These situations don't come up very often, but I do want you to know that there can be times when "always having an alternate" can be inconvenient, but that I still think it is worth observing because it is very hard to run out of gas, or even come close to running out of gas, if you make a point of always having the alternate. When this situation does occur, the solution is very simple: plan on a stop. Then the trip can be flown with an alternate for each leg.

HAZARDOUS WEATHER

If you have received a good weather briefing, have filed IFR, have prepared a detailed flight log, and have designated and fueled for a conservative and legal alternate, about the only thing left that can get you is hazardous weather, and that's what the "Hs" stands for in BILAHs—hazardous weather.

Information on hazardous weather comes during the first part of the FAA Standard Weather Briefing (adverse conditions), and so, in a way, this fifth step is redundant. I could say to make sure and check for hazardous weather during the briefing and let it go at that, but I like to keep the two separate for a couple of reasons. First, if you are going to file IFR anyway, then the main purpose of the first step in the strategy—the briefing—is to obtain the information you need to flight plan properly: what are the winds aloft and what kind of ground speed can I expect, what are the ceilings along the route of flight, will there possibly be weather-related delays that I should fuel for, can I expect an instrument approach at my destination, what's a good alternate? Secondly, the main reason for the last step is to ensure that the existence of any hazardous weather has been specifically considered and has not been lost in the general weather briefing.

So, even though in theory checking and developing contingency plans for any hazardous weather should be taken care of automatically in simply asking for a standard weather briefing, I think hazardous weather is important enough to warrant a final check at the end as a part of a complete, safe flying strategy.

What is "hazardous weather" and what kind of contingency planning can be done to avoid it? As I see it, hazardous weather falls into one of five major categories: thunderstorms, ice, freezing rain, snow, and severe turbulence. There is an inevitable overlapping of categories—ice and freezing rain, thunderstorms and turbulence—but each hazard is worth considering separately.

STRATEGY

Thunderstorms

Thunderstorms are the ultimate "real time" events, and real time events are, unfortunately, what aviation has the most trouble with. (Even terminal sequences—current weather—are at least 15 minutes old by the time you get them.) A thunderstorm can come and go in 30 minutes, so unless the information about it is real time, it is fairly useless. The real time radar repeaters that certain flight service stations have cover such a large area that what you are looking at are not necessarily individual thunderstorms, but are more likely *areas* of thunderstorms. This is good information, but unfortunately it isn't detailed enough to use for an area penetration.

The simple, sad fact is that if you don't have airborne radar (or possibly a lightning detecting system, but I have no experience with that system), all you can do is avoid areas of thunderstorm activity by a wide margin. The airlines, for instance, aren't allowed to dispatch an airplane without a working radar unless there is virtually no chance of thunderstorm activity along the entire route of flight. If you don't have radar or any other source of thunderstorm detection equipment, the purpose of this part of the briefing is simply to inform you what route you must take to avoid all areas of possible or known thunderstorm activity.

If you do have airborne radar, then you must take the information you can get from the briefing on the size and type of thunderstorms forecast, compare it with the capability of the equipment, and make a decision. All radars are not equally capable. A radar's ability to see ahead, and its accuracy and detail in displaying the information it receives, is a direct function of its power, the size and type of antenna, and the sophistication of its circuitry. The more capable the radar, the more direct the route of flight can be.

I'm not going to try to tell you how to use your radar in this book. There are several very good ground schools you can attend on radar, and a lot has been written already on the subject. But I can say that the smaller and simpler radars typically found on heavy singles and light twins can only help in avoiding general *areas* of thunderstorm activity. They don't have the power to see far enough ahead to give you a reliable path *through* an area of thunderstorms. With increasing power and sensitivity, radar can do a good job of showing ways in and around individual cells, but you usually have to fly in some fairly heavy iron before you find radars with this kind of capability.

I wish I could tell you more about thunderstorms and radar. I'd like to explain when you can fly with thunderstorms in the forecast and when you can't, when you can trust a "hole" to be safe and when you can't, how much power it takes to "see" embedded thunderstorms in a bunch of rain, and so on. In short, I wish I could give you a system for dealing with thunderstorms, but I can't.

There are just too many variables and, at any rate, neither conventional radar nor lightning detectors tell you directly what you really want to know, which is "Where is the turbulence?" Radar can point to it by inference—where there is heavy rain there often is severe turbulence—but it can't see turbulence directly.

The only device that detects turbulence directly is *doppler* radar. The Doppler Principle in meteorology says that raindrops moving away will appear to reflect energy of a

Hazardous Weather

lower wavelength than raindrops moving forward. Because turbulence is merely air that is going in all directions at rapidly changing rates, applying the Doppler Principle to radar reflections (via computer, of course, which looks at the frequencies of all raindrops and decides which direction and what speed they are all going), gives a picture not just of raindrops, but a picture of the turbulence. (But doppler can't see clear air turbulence. Doppler is still based on radio energy and needs moisture—or something more substantial than air—to bounce off.)

Doppler technology is just beginning to be incorporated into airborne radar but it is still not terribly common. Nonetheless, I think applying Doppler Principles to airborne radar is the biggest advance in thunderstorm avoidance since the radar itself. It will probably be awhile before it is available and affordable for general aviation, but it should be worth the wait.

In the meantime, if you don't have radar capability, stay completely away from thunderstorms. If you have limited radar capability, use it to fly around areas of thunderstorm activity. If you have a lot of radar capability, use it to stay 10 miles away from small cells, and 20 miles away from big cells. Conventional airplanes have about as much business inside thunderstorms as butterflies do in wind tunnels.

Ice

Ice can be a real problem for any airplane, but particularly for aircraft with limited deice/anti-ice capability. Fortunately, information on icing conditions is one area where a lot of fairly useful information is normally available, and the time factor—the speed with which icing conditions come and go—is not nearly so critical as it is with thunderstorms. The weather briefer will be able to provide information about where the icing level is, what type of icing is forecast (clear, rime, or mixed) and the amount expected (trace, light, moderate, or severe). He will be able to pass on any pilot reports of actual icing, and he should point out any stations reporting freezing precipitation of any kind along the route of flight.

If an aircraft is prohibited from operating in known-icing conditions, then any pilot or station report of actual icing conditions precludes operating in that area. This is a very gray area as far as legality is concerned, complete with holes large enough to fly a 747 through. How old is the information? When is the information old enough you can disregard it? How big is an "area?"

But, of course, the idea isn't to find loopholes in the law. The idea is use the briefing to avoid icing conditions if the aircraft is not approved for flight in icing conditions, and to minimize your exposure to ice if it is approved. Whether icing is reported or simply forecast is an academic question. The purpose of the weather briefing is to provide information for flight planning purposes, and for planning purposes you have to assume that the forecast is correct.

Does this mean you have to automatically cancel every time someone says "ice?" No, of course not. The purpose of the briefing is to help you find a way the flight can be safely

193

STRATEGY

accomplished, and the purpose of this last step, the check on hazardous weather, is simply to make sure you have done that.

One way to avoid ice is to simply stay below the icing level if that is possible, which often occurs in the fall and spring, but usually not in the winter. If an aircraft is approved for flight in known icing conditions, you can often climb on top, or least you can often climb high enough that it is so cold ice is not a problem. Colder is generally better when it comes to avoiding ice. That doesn't make a lot of obvious sense, but that's how the mechanics of ice formation work. Also, sometimes by flight planning around a mountainous area, or by avoiding the lee side of a large body of water, the icing area can be avoided.

With approval for flight in known-icing, you have the option of flight planning into areas of light or moderate icing. This really isn't as reckless as it might seem if you have provided for an "escape route." It is pretty much standard operating procedure to include the possibility of light to moderate rime icing in all winter forecasts, just to be on the safe side.

You usually can find a little ice at one altitude or another if you try hard enough, but it usually can be handled with normal deicing equipment. Every now and then, for various unpredictable reasons, light to moderate icing becomes severe—well beyond the capability of the airplane to shed or carry—and when that happens it usually happens very quickly and without too much warning.

When it does happen, you have to have an "out," an alternate course of action. The most common "out" is a known area of relatively warm air nearby. Sometimes the warm air will be behind you, and sometimes below you. Occasionally, when the tops are low, or the temperature is unusually cold, or when there is an inversion of temperatures (warm air *above*), your "out" will be to climb. The last choice is a lateral move *away* from the icing area. This works only when the icing is a relatively local phenomenon, such as around a lake, in the mountains, or flying east-west along the southern edge of an icing area. Trying to fly away from an icing area is risky because you sometimes have to fly a long way to escape to noicing conditions.

This is why the first step, the weather briefing, and this last step, specifically dealing with any hazardous weather, are so important: you want to know ahead of time what the best move is if the light to moderate icing turns out to be severe. Any time you elect to test the capability of your aircraft to handle known or forecast icing conditions, it is important that you find out in the briefing where the closest ice-free air is. If the cold air goes all the way to the ground, or the tops are too high to climb above, or the nearest area not forecasting ice is hundreds of miles away—if in other words, you don't have any "outs"—then you might have picked the wrong day to want to go somewhere.

Freezing Rain

One situation that is virtually an automatic "no go" for all pilots flying any kind of airplane is freezing rain, or its cousin, ice pellets—what you and I call "sleet." As soon as you see, hear, or feel freezing rain outside, or as soon as the briefer says anything about freezing rain at your destination, just put your stuff away and ask what the long term out-

Hazardous Weather

look is. The reason freezing rain is hazardous for all aircraft is because deice and anti-ice systems are meant to protect the impact areas of critical components and surfaces, like the leading edge of the wing and the windshield and pitot tube, but they are not capable of clearing an entire airplane of ice and that is what happens in freezing rain. The ice that forms quickly adds weight to the aircraft, changes the aerodynamic characteristics, and diminishes the lifting ability of the wing by changing its shape. Freezing rain is always to be avoided.

If you find yourself in freezing rain en route, then you clearly have missed the chance to avoid it and now must do something to get out of it, and you must do it quickly. The key here is to understand what the circumstances are that cause freezing rain; freezing rain is ordinary rain that has fallen into cold, below freezing air. This means that there is always warm air *above* any area of freezing rain. (The difference between freezing rain and snow is that freezing rain starts out in warm air and falls into cold air; snow forms in cold air and stays in cold air, or, if it does fall into warm air, turns into either ice pellets or cold rain.)

When inadvertently flying in freezing rain, the solution is to climb into the warmer air above if possible and do it immediately, at the first sign of freezing rain. (Normally the first sign will be clear, frozen water on the windshield.) Tell the controller you are experiencing freezing rain and *require* a higher altitude.

Usually the warm air will be just a couple of thousand feet higher. If he can't give you higher, ask for a "180" back to nonfreezing rain conditions. If he can't give you that, ask for an approach to the nearest airport and stress the urgency. You need to get on ground fast, while control is still possible.

If you don't have much climb capability when you first encounter the freezing rain, then you should not try to climb into the warmer air because that might make matters even worse if you don't make it. Your only choice then is the "180." A quick way to tell how much climb capability remains is to apply climb power, wait for the airplane to accelerate, and then compare your indicated cruise speed to the best rate-of-climb speed—V_y. That difference represents excess power available for climbing. If there isn't a pretty good margin between the two—30 or 40 knots—don't try to out-climb the freezing rain.

Just as with unexpected severe icing, encountering freezing rain en route is usually a fairly easy problem to solve if you do something *right away,* but don't fool around waiting for conditions to get better; it will get worse. The idea, obviously, is to avoid getting into conditions of freezing rain in the first place, but if you do get into the conditions, *do* something about it, don't just sit there.

Snow

Snow is sneaky because it can get you in so many different ways, and sometimes it doesn't bother you at all. Generally, the colder it is, snow is less of a problem. Cold snow is dry snow, which means it blows around a lot, sometimes causing visibility problems, but it won't stick to the airplane flying through, and snow won't stick to the airplane on the ground if the metal is below freezing. (This means cold-soaking the airframe prior to

Strategy

bringing it out of the hangar. If you bring a warm airplane out of the hangar into the snow, the snow will melt on contact, and then freeze, and you will have to take it back inside and start over again.) So cold is generally good when it comes to snow.

If it is snowing at the time of departure and the snow isn't dry enough (cold enough) to blow off by the time you get to the runup area, the snow isn't going to come off on the takeoff roll either. If you try to depart with snow on the wings, you will not only be illegal, you will also be stupid.

Here's a short review of advance aerodynamics: The reason an airplane flies is because it has a wing. What makes a wing a wing, and not just a good place to put the fuel, is its shape. If you change the shape in any way, such as putting snow all over it, you no longer have a wing, although you do still have a big place to put a lot of fuel. That is not a very good combination.

So, if you get to the runway and there is snow on the wing, you're going to have to go back and remove the snow, and then you're going to have to deice the airplane to prevent the same thing from happening again. (Technically this is anti-icing—the *prevention* of ice formation; deicing means the *removal* of ice, but everyone calls it deicing anyway.)

Never assume that the airflow on takeoff will blow the snow off. Some of the snow might blow off—maybe the top inch or so, and some of the snow along parts of the fuselage—but the snow close to the wing probably won't come off. I know that doesn't make sense; you would think that at 100 knots or more, anything would blow off, but it's not true. The air in the layer next to the wing is actually still and doesn't move. (That's why dust and dirt don't blow off the wing either.) Therefore, if there is any snow left on the wing prior to takeoff, you have no choice but to return and remove it.

The correct way to deice an airplane is to use heated, full-strength glycol. Even a good soaking with hot glycol will only keep heavy snow off for 10 or 15 minutes, and anything less—cold glycol or a diluted solution—will reduce that interval even less. Hot glycol is expensive, but this is not the place to compromise on materials.

While the airplane is being deiced, work on a plan to get from the hangar to the runway as quickly as possible. If the airport isn't busy, you can call the tower and tell them you want to stay in the hangar until released by ATC. Get everyone in the airplane, do as much of the pretakeoff checklist as you can, and as soon as you get the release, have somebody tow you out, unhook and remove the tow bar (at least one airplane has become airborne, briefly, with a tow bar attached), complete the checklist, and move smartly to the end of the runway. If all goes well, you should be out in the open for only a few minutes prior to takeoff, and the wings should stay snow free.

If not, then you have lost the gamble and are going to have to go back to the hangar. Nice try, though. If the tower can't approve waiting in the hangar for the release, you can still try, but the chances of success go down. At a busy airport with a long line waiting to go, you probably cannot get any consideration. If it is snowing hard, save money and wait for another day. Life's too short as it is. Go find a nice fireplace and a good book and enjoy the storm.

During the winter months the briefer might have information on runway and taxiway conditions at the intended destination in the NOTAMs. The best and most current information can only come from calling ahead and talking to the people actually on the scene. Ask how the plowing is going, what percentage of the runway surface is bare, how high the snowbanks are, whether the runway is sanded, whether the sand was put on hot and is sticking or was put on cold and is rapidly blowing away, what the ramp looks like, and so on. If refueling is a factor, ask about the fuel truck or pumps because they love to break down in snowstorms. These are all important factors that in all likelihood can only be determined by a direct phone call, but the weather briefing is a starting point that might reveal through a NOTAM that the airport is closed and that's that.

Turbulence

Turbulence isn't always hazardous, but it is always uncomfortable and most pilots like to avoid it as much as possible for that reason alone. When turbulence is a function of thunderstorm activity, clear air turbulence, or rotor turbulence under a standing wave, then it can definitely be hazardous. Any turbulence labeled moderate to severe or worse, regardless of the cause, should be considered hazardous.

During the weather briefing you want to note not only where the *areas* of turbulence are (in case these areas can be avoided), but also how *high* the turbulence is forecast to go, in case you can climb above it. If the turbulence can't be avoided, think about slowing down to maneuvering speed. This will be a little more comfortable and much easier on the airplane. In terms of turbulence, the purpose of this last check is simply to make sure you have given some consideration to hazardous turbulence, and have done something specific to avoid it.

CONCLUSION

Arnold Palmer, in describing the five key components common to all great golf swings, was careful not to say that those five items were all that mattered to play great golf. But he did say that if those five items are observed, the other components of a good golf swing would come into play.

That's what I hope this five-part strategy for safe cross-country flying does, too. I am not saying that these five things are all that matters in flying, but I am saying that if you follow this five-part strategy diligently and thoroughly, you can hardly miss not taking care of everything else that matters too:

Briefing: Get a complete weather briefing.
IFR: File IFR.
Log: Prepare and maintain a good flight log.
Alternate: Always have an alternate.
Hazardous conditions: Always specifically check for hazardous conditions.

I just don't see how you can go wrong if you consistently observe these rules, but I can see many ways to go wrong if you don't. Just like my golf swing . . .

13
Flight Levels

Fly Like a Pro, MY FIRST BOOK AND UPON WHICH THIS BOOK IS BASED, CONCLUDED with a chapter on flying professionally (flying as a professional, as opposed to flying *like* a professional). Since that book was written, professional aviation has changed so significantly and so completely (largely as a result of deregulation), that not only is that chapter out of date, but the very idea of being able to write something about professional aviation—something that won't be obsolete before it even reaches print—is also obsolete.

You might still (I hope) enjoy going back to that book and reading about some of my experiences as a corporate pilot, and much of what I said in that book about getting started in aviation is still relevant and always will be. However I can no longer describe exactly how to break into professional aviation, and I certainly can't tell you what form the typical aviation career will take in the future.

Professional aviation is changing rapidly and generally for the better—the opportunities are enormous—but your guess is as good as mine as to where we go from here. (If you are truly interested in a career in aviation, I recommend joining FAPA—Future Aviation Professionals of America: 1-(800)-JET-JOBS. FAPA is the single best source of information on jobs and aviation careers at this time, and has been for several years. Good luck.)

What I have learned since that first book (after changing from corporate to airline flying, and with time out to write a couple of other books), is perhaps more important anyway. What I have learned, or more accurately, been forced to relearn, is that flying is based not on an infinite series of facts and skills, but on a core of knowledge and skills, a core that is itself constant; I have learned that an infinite number of additional layers of skill, knowledge, and wisdom can be added to that core, and that while each additional

FLIGHT LEVELS

layer is important to consolidating, amplifying, and solidifying that core, in the end it is the core itself that matters.

This might sound like an obscure point, but I think it is, in fact, quite significant, and I'd like to see if I can't explain why I think it is so in the rest of this last chapter. It's going to take a little jumping around to do that, but stick with me.

SERIAL LEARNING

Aviation with a solid and consistent core that is common to all operations and all levels of flying is not the way pilots usually think about aviation. At least, it has not been the way I have always thought about aviation. I have always thought, unconsciously at least, that in attempting to master aviation, first I would learn "A," and then I would learn "B," and then I would learn "C," and if I kept doing that, eventually I would get to some practical end.

I knew that I could never actually learn everything in aviation, but I felt that I could learn everything, or nearly everything, that I actually needed to know, and that after a certain point I could concentrate on retention and refinement, not on further knowledge and skill. I felt, in short, that having learned A, and having then gone on to learn B and after that C, that eventually I would get to Z, or at least Y. I know that not only do you never get to Z, but that you don't even go from A to B to C—you just keep adding to A.

BACKGROUND

People sometimes ask how I got into aviation, and I usually explain that I started taking lessons when I was in the Army at the post flying club, and that after I got out of the Army I used my GI Bill's benefits (after a couple of forays into other careers) to get my commercial and instructor's ratings, and that I then worked as a flight instructor for several years, did some multiengine charter, was hired by an aircraft management company as a copilot and eventually became a Citation captain, then a corporate job as a Citation and later Falcon 20 captain, wrote books on flying, and that I am now flying L-1011s internationally as a first officer.

That's my professional life, and describing it that way makes it sound like I learned to fly at some point, more or less proved that I could fly by eventually becoming a jet captain, and now I simply *fly* and that is my job. Too bad it's not that simple. Actually, I don't mean that. Thank goodness it's not that simple. To explain what I meant by *that*, I have to go back even farther.

DIFFERENCES

There is nothing fundamentally different between a Cherokee 140 and an L-1011, nor is there anything fundamentally different between what a person needs to know to fly either one. I know you probably don't believe that, but it's true. I didn't used to believe it either. There are, of course, numerous differences in the details between flying a Cherokee and flying an L-1011, but there is nothing fundamentally different between the way the two are flown.

They both have flight controls and flight surfaces, and they both have a source of thrust, and the fact that one set of flight surfaces is connected by cables to the control column while the other set is connected by hydraulics, and the fact that one airplane has a single engine for a power source while with the other airplane power is divided into three separate parts means there are significant differences in aircraft systems and in aircraft control when a power source fails. However, there is nothing fundamentally different between the two airplanes. Regardless of the kind of airplane you normally fly, I could put you into an L-1011 and you could fly it, because when you push on the control column of a 1011, the nose drops just like it does in a trainer. It doesn't matter that the 1011 has a great big nose way up in front of a big long airplane, nor does it matter that hydraulic pressure has been substituted for muscle power to move the elevator, the nose simply goes down and that is that.

Learning to fly an L-1011 (as just one example) is not something new and different from flying a Cherokee 140: it is the same thing only different. And that is what I mean by flying being layers of learning built upon a core of skill and knowledge rather than a series of facts and skills.

Perhaps I can describe this core and then take you through some of the layers as I see them. I want you to see that what you and I do is not different, that we all do the same thing only in different airplanes, that you too have a core of knowledge and skill with many layers above that core, that it is always possible to add another layer, no matter how experienced the pilot, and most importantly, I want you to know that when you quit adding layers, you have ceased being a pilot, because, in the final analysis, learning is what flying is all about. Without learning, a pilot is simply a machine, and all machines eventually break.

THE CORE

The core, for me, upon which all further layers were added, consisted of the basic flying licenses and ratings that I first obtained as a hopeful professional pilot: a private pilot's license with instrument rating and later a commercial license with multiengine rating. This core of knowledge and skill didn't come easy, nor was it without moments of frustration and disappointment. It took several years overall to achieve and cost several thousand dollars in flight time and instruction to accomplish, but eventually I was able to acquire the basic skills and knowledge necessary to become a reasonably competent pilot, and that core still forms the basis for everything I know and do in aviation today.

Those licenses and ratings, in fact, legally enable me to do what I do today, but more importantly, as a practical matter, what I learned to do then in Cherokees and Twin Comanches is still exactly the same thing I do now: taxi, takeoff, climb, cruise, descend, approach, and land airplanes.

The difference, though, between what I thought then and what I think now, is that I once thought that having finally achieved all of these ratings, I had learned everything I really needed to know about flying airplanes, and in the most fundamental sense I was right; I did not know the details and specifics of flying other airplanes in other commercial operations, which left an enormous amount that I did not know. I can see, looking back, that I was intuitively correct about my having achieved a core upon which everything else in aviation would be built.

INSTRUCTION

I was correct about achieving a core of aviation skill and knowledge, but I was wrong in thinking I had mastered that core—I was a long ways from mastery, as the next step in my learning process demonstrated. The next step, for me, was getting my flight instructor ratings so I could teach flying, make some money (a double plus after having had to spend so much money), and get some practical experience (a nice way to say "built some time") so I could go on in aviation.

(I don't want to downgrade the importance of flight instruction or flight instructors. I loved being a flight instructor, and I think it is one of the most satisfying and important jobs in aviation, but I couldn't live on a flight instructor's pay, and I was, I readily admit, anxious to fly larger, faster, and more-complex airplanes, just like everybody else, and to do that I needed flight time.)

I thought at the time that becoming a flight instructor meant learning to fly the airplane from the right seat, learning a flight curricula, learning to troubleshoot student mistakes, and learning how to transfer what I knew to students who wanted to learn. I was right because learning to be a flight instructor is indeed all those things, however learning to be a flight instructor is mostly learning to fly again.

You simply cannot be a credible instructor unless you can fly the airplane quite well yourself. If the instructor can't do a maneuver, and do it well, how can he expect the student to do it? Instructors have to be able to demonstrate the maneuvers, they have to know the difference between a maneuver done well and a maneuver done imperfectly, and after critiquing a student's maneuver, they have to be able to show how it should be done. Becoming a flight instructor forced me to be the pilot I was supposed to be anyway.

This illustrates one thing about flying—maybe about a lot of things—that I have discovered over the years, and that is, to reach one level of expertise, you have to move past to the next level. I didn't really learn to fly until I became a flight instructor and was forced to do it correctly (and I didn't even start to become a commercial pilot until I had an airline transport pilot rating, but that's getting ahead of myself). This isn't to say that the core of flying knowledge and skill that I talked about is not truly a core, but it is to say that the core is not truly activated, or validated, or legitimatized until it has been hardened by additional experience and training.

COMMERCIAL FLYING

After I had instructed for awhile, I started to have chances to do some "real" commercial flying, first as a VFR-only single-engine charter pilot (I could file IFR, but the weather had to be VFR to be legal for single-engine charter), and later in IFR multiengine charter. I assumed at this point, again incorrectly, that because I had so many ratings in the first place, and because I was certified to teach people to get those same ratings in the second place, that, clearly, I had to be a fully capable and competent commercial pilot myself.

I was wrong. I wasn't truly a fully capable and competent commercial pilot yet, but I wasn't dangerous either. I just didn't know what I really needed to know to do the job. As we all know, being able to teach and being able to do are two different things, and, besides, the reality of actually taking many different people and things to a variety of different places under all types of conditions—often on short notice and in airplanes that were not as perfectly maintained or equipped as they could or should have been—is much different from the near-classroom conditions of flight instructing.

I did many things in the "early" days by myself and without the benefit of coaching or assistance, and while they worked, I was also doing a lot of things incorrectly, or at any rate, if not incorrectly, certainly inadequately. I never knew that I didn't know enough. I realized that my flights didn't go as routinely and smoothly as it seemed they should; I didn't know why because I was always flying by myself and nobody was there to tell me anything different.

THE RIGHT SEAT

I really didn't learn how to become a true professional pilot—someone who not only gets paid to fly people around, but someone who *deserves* to get paid to fly people around—until I was hired as a copilot on heavy twins and light jets. It dawned on me what a tremendous amount of information had to be absorbed to even begin to be a competent professional pilot. After all, years before, I thought I became a pilot when I earned a private pilot's license, and then I knew that I was a real pilot with an instrument rating, and then further realizations with each additional license and rating after that, and here I was, in theory fully qualified to do what I had been hired to do, but in fact almost having to learn everything about aviation all over again.

In retrospect it is probably safe to say that I learned as much as a copilot flying with experienced captains on corporate jets as I did in all of my flying before that. That probably sounds like an exaggeration, but it's not, not by much anyway. Flying as a copilot in Cessna 421s and Citations and later on a Hawker-Siddely I learned how transport flying is done on a practical and safe daily basis. In fact, most things explained in this book I learned as a copilot, not as a captain. Captains are too busy making sure everything goes right (which includes looking after new copilots) to learn much. Copilots learn things, and if they don't, they don't deserve to become captains.

THE LEFT SEAT

Still, getting into the "left seat"—the captain's seat—is a big step. Being a captain means being responsible for the entire flight, which means that most of your time will be

FLIGHT LEVELS

spent checking and reviewing, not learning (you will, of course, learn things anyway, but that is a secondary effect). For me the step up to captain was analogous to the step up from private and commercial pilot to flight instructor: I didn't really learn how to handle airplanes until I tried to teach other people how to do so, and I didn't really master being in control of a flight until I found myself in the left seat of a transport category airplane with a copilot in the right seat and several rows of passengers behind me.

It's a great feeling to be in charge in a situation like that, and tremendously satisfying and rewarding when all goes well, but it is also a tremendous burden of responsibility to carry, one that forces you to be sure you know what you "think" you know, and exposes very quickly what you don't know but "think" you do.

THE RIGHT SEAT ... AGAIN

Logically, that should be the end of the story—to paraphrase the old joke about a college education, "Twintee yers a go I cudn't spel 'captain' and now i are one"—but it is not. In changing from corporate to airline flying I also changed seats again, back to the right seat. But moving back to the right seat seems a small price to pay for the chance to do this flying—in fact, I wouldn't want it any other way (FIG. 13-1).

The flying is great, the airplane is the highest performing and most capable airplane I have ever flown, we have flight attendants, flight engineers, dispatchers, and maintenance reps to take care of most of the nonflying duties, and it is fun flying from the right side

Fig. 13-1. *The right seat—in this case the right seat of an Lockheed 1011—is where copilots learn to be captains.*

again. But it does mean that I have to readjust my thinking from supervising and teaching back to helping and learning, which isn't so bad.

It is like a vacation, or getting to go back to school after awhile out on the road or on the line—a privilege, of sorts. Instead of having to be in charge and worry about every little thing, I am free to look around, to ask questions, to observe, to experiment without jeopardizing safety, to study, and to practice. I hope, of course, that some day in the not too distant future I will again be a captain, but I am happy at this point to have the opportunity to learn how to be an L-1011 captain from the right seat.

Perhaps what I am most struck by at this point in the learning process is that what I am doing is different in detail from what I did before, but not different in substance from anything I have done before. All airplanes are fundamentally alike and all have to be operated in essentially the same way. The L-1011 might have two five-inch attitude indicators instead of a single three-inch artificial horizon, but there is nothing fundamentally different between the information presented by either. Dispatchers and a computer do the fuel and flight planning, but how it gets done is irrelevant because the trips still have to be flight planned, and somebody still has to keep track of the flight log in-flight.

The L-1011 might have three engines to launch itself up into the air instead of one or two, and those engines might be jet engines large enough to stand up in instead of smaller, reciprocating engines driving propellers, but an airplane doesn't know or care how many engines are necessary to push or pull it or what form those engines take, so long as something pushes or pulls it.

(There are substantial differences between the way a jet engine responds and the way a reciprocating engine, or any engine with propellers, responds, and there are, of course, substantial differences in the way these aircraft are handled when a power unit fails, but in the final analysis these are differences in details, not in fundamentals.)

In short, what I am learning now might seem like something new, and in certain ultimately unimportant ways it is, but it isn't actually—it is the same thing only different. In the end, the layers of learning are translucent, and when you look down into the core, what you see is the same thing—boxes within boxes within more boxes, all alike.

PILOT-IN-COMMAND

You might not have the opportunity to get additional ratings, or to fly from the right seat of a transport category aircraft, or to fly commercially or internationally—to add layers of learning the same way I have been lucky enough to—but you can still add layers. Those layers can be added in the form of additional experience, additional training, additional study, and, most importantly, additional thinking about what this game is all about, what matters and what does not.

It is very important to remember that no matter how many layers are added, what lies at the core is the same for every pilot and that is *competence in handling the airplane under a variety of different conditions in order to safely and routinely take you and your passengers from A to B*. The key is the core, and the more solid and stable you can make that core, the better.

FLIGHT LEVELS

Different levels of flight operations exist, but a single-engine airplane flown at 6,000 feet by a private pilot who has supervised his own maintenance, planned and dispatched his own flight, filed his own flight plan, and done everything that needs to be done in the airplane without the benefit of a copilot, is not different in any significant way from a three-engine jumbo jet flown at Flight Level 370 with a crew of 11 and a small army of support troops behind it. Each flight is a way of taking people and things from A to B, nothing more and nothing less.

The general aviation private pilot is not at the bottom of the aviation ladder, he is at the top. If that is you, my advice is to take pride in what you do, and do everything you can to do that job correctly to warrant that pride.

Never forget that you have the most important, the most difficult, and the most prestigious position in aviation: pilot-in-command.

Index

A

abnormal situations
  checklists for, 154
  emergency vs. 144
  flaps stuck or split, 156
  gear problems, 158
  generator failure, 155
  heater inoperative, 158
  hydraulic pressure loss, 156
  inoperative engine, 157
  pitot heat inoperative, 157
  precautionary shutdown, 157
  professionalism in, 158-159, 158
  static system clogged, 158
  vacuum pump failure, 158
  warning lights on, 158
adverse conditions, weather briefings, 116
aileron, trimming, 3
air route traffic control center (ARTCC), 29
airspeed control, 30-33, 48
affirmative, 134
Airman's Information Manual, 14, 115
Airport Advisory Service (AAS), 19
Airport/Facility Directory, 28
airport radar service area (ARSA), 28
airspeed control, 1, 12, 13, 69

alternate airports, 35-37, 124-125, 190-191
altitude control, 1, 3-5, 13
  circling approach, 70
  flight plans and, 30-33
  leveling off, 10
  minimum en route (MEA), 31
  oxygen, 31-32
  porpoising, 8
  steep turns, 8
  weather factors: ice, turbulence, etc., 32-33
  wind speed and direction, 32
anxiety, 106
approaches, 63-81
  circling, 7, 64-74
  holding patterns, 2, 15
  instrument landing system (ILS), 2, 78-80
  LOC, 75
  NDB, 75, 77-78
  nonprecision, 74-77
  precision approach radar (PAR), 80-81
  SDF, 75
  shooting, 6, 13
  vectored, 75-76
  VOR, 75
area charts, 29

area navigation (RNAV), 17
ATC delays, weather briefings, 124
attitude control, steep turns, 8
autopilots, 11-13, 17
Aviator's Guide to Flight Planning, The, 25
Aviator's Guide to Modern Navigation, The, 81
Aviation Weather, 112

B

back pressure, 8
backup plans, 83-84
briefings (see weather briefings)

C

cabin smoke or fire, 148
calling tower, 135
CAT III ILS, 81
ceilings, 35, 107-108
circling approaches,
   altitude control, 70
   circling patterns, 66, 70-74
   descent to MDA, 67
   heading vs. runway, OBS setting, 70, 71
   legal limitations, 67-69
   optimum conditions, 64-67
   regulations on, 68
   visibility restrictions, 68
   VOR/DME, 65
clarity of communications, 130
clearances, 22-23, 135
climb-out, 7
climbs, 1, 8-11, 93-94
copilots, 203-206
commercial aviation, 202
communications, 1, 127-140
   calling tower, 135
   clarity in, 130
   clearances, 135
   controllers and, 128-129
   data links, 129
   flight service stations, 17-19
   frequency management, 137
   key words in, 133-135
   language, 132-133
   limitations of radio, 129
   memory tricks for, 138-140
   problems in, 130
   radio procedures, 140
   sidetones and echoes, 131
   simultaneous transmissions, 140
   squelch, 138
   verification, 130, 135
   volume, 138
computer weather briefings, 126
contingency plans, 83-84
convection, 32
course change en route, 7
cross-country flight, limitations, pilot, 103
cruise control, 45-61
   en route options, 46
   en route performance, 45
   high-speed, 54-55
   leaning and detonation in engine, 49, 50
   long-range, 46-49
   low-power, 49, 51, 52
   lowest cost (LCC), 57-60
   maximum, 55-57
   mileage, 45, 46
   normal, 52-54
   very-low-power, 49, 51
current conditions, weather briefings, 116
customs, 18

D

data link communications, 129
decision making, judgment, 174
delays, ATC, weather briefings, 124
depressurization, rapid, 151
descent, 1, 7, 11, 12, 152
destination airport, 27-28
   alternate, 35-37
   ceiling, 35
   visibility, 35
destination forecast, weather briefings, 120
detonation, 49-50
ditching, overwater flight, 87, 150
dives, 11

E

electrical fire, 148
elevator, trim, 4
emergency situations, (*see also abnormal situations*), 2, 9, 19, 141-159
   abnormal conditions vs., 144
   cabin smoke or fire, 148
   checklists for, 145
   critical elements to remember, 144-145

cruise control, maximum cruise, 55-57
depressurization, rapid, 151
descent procedures, 152
ditching, 150
electrical fire, 148
engine failure, 146
engine fire, 145
forced landings, 149-150
generator/alternator failures, complete, 153
inhospitable terrain, 88-89
nonelectrical fire, 149
problem solving, checklist for, 142-144
professionalism in, 158-159
propeller overspeed, 147
spins, 154
trim runaway, 149
en route charts, 29, 118
engine
   failure, 146
   fire, 145
   inoperative, 157
   limitations, aircraft, 85-86
   precautionary shutdown, 157
engine-out climb gradients, 93-94
experience, 179

F

fires, 145-149
flaps stuck or split, 156
flight following, 2
flight levels, 199-206
flight logs, 39-41, 189
flight manual, aircraft, 1, 19
flight plans, 1, 2, 18, 25-43
   airspeed, 30-33
   alternate airports, 35-37
   altitude, 27, 30-33
   cruise power settings, 31
   destination airport, 27-28
   deviations from route, 29
   distance, 30
   flight logs, 39-41
   fuel requirements, 26, 27, 30-31
   high level, 25
   IFR, 25, 26
   international, 25
   major components of, 26-27
   maximum range trips, 41
   power selection, 32
   refueling aircraft, 38-39

reserve fuel, 37-38
route, 26, 27
routes, preferred, 28-29
time and fuel estimation, 33-35
VFR, 25, 26
weight and balance, 41-43
winds, altitude selection and, 32
flight service stations (FSS), 1, 17-19
flight tests, 182
flight training, 168-169
Fly Like A Pro, 199
forced landings, 149-150
forecasting, 118-120
freezing rain, 194-195
frequency management in communications, 137
frequent flying, maintaining proficiency through, 165-167
fuel requirements, 26, 27, 39-40
   cruise control, 45, 2
   flight plans, 30-31, 33-35
   maximum range trips, 41
   refueling aircraft, 38-39
   reserve fuel, 37-38
   weight and balance calculations, 41-43
   winds and, 34

G

generator/alternator failure, 153-155

H

headings, 3, 6, 17
heater inoperative, 158
high-speed cruise control, 54-55
holding patterns, 2, 13-15
hydraulic pressure loss, 156

I

ice, 32, 193-194
IFR, 2
   flight plans, 26
   limitations, pilot, 102, 104-105
   single-engine, limitations, aircraft, 90-91, 95
   strategy for, 187-189
   takeoff, 95
inertial navigation, 17
information-on-request, weather briefings, 124
inhospitable terrain, limitations, aircraft, 88-89
insight, judgment and, 174

instinct, limitations, pilot and, 100
instruction
   flight training, 168-169
   learning and judgment, 174
   multiengine flight training, 169
   procedures training, 168
   professional aviation, 202
   recurrency training, 164
   simulator training, 165
   single-engine flight training, 169
instrument landing system (ILS)
   approaches, 13, 78-80, 178

J

jets, limitations, aircraft, 97-98
judgment, 173-182
   conservatism, 181
   decision making, 174
   essential elements of, 176
   flight tests, 182
   gray areas of, 175
   hard calls, 180
   insight and, 174
   knowledge, experience, skill, personal elements, 179
   learning and, 174
   skill or lack thereof, 177

K

key words in communications, 133-135
knowledge, 179

L

landing gear problems, 158
landings, 11, 89-90, 149-150
language of communications, 132-133
leaning engine, 49-50
letdowns, nonprecision approaches as, 75
leveling off, 1, 10-11
lift over drag (L/D) speed, 48
limitations, aircraft, 1, 19-21, 83-98
   basic limitations, 84-85
   contingencies and backups, 83-84
   emergency landings, 88-89
   engine reliability, 85-86
   engine-out climb gradients, 93-94
   inhospitable terrain, 88-89
   jets, 97-98
   multiengine takeoff performance, 91-93
   overwater flight, 86-88
   single-engine IFR, 90-91, 95
   single-engine night flight, 91
   takeoff alternatives, 93
   turboprops, 96-97
   urban area takeoff and landing, 89-90
   weather limitations and, 98
limitations, pilot, 99-109
   anxiety, 106
   day trips, 102-103
   extended cross-country flight, 103
   IFR benefits and, 104-105
   instinct, 100
   instrument training, 102
   low ceilings, 107-108
   new IFR pilots, 105-106
   new pilots, 106
   new private pilots, 101-102
   privileges vs. limits, 99-100
   systematic approach to, 100-101
   upgrading performance, 107
   VFR, 103-104
   visibility, 107-108
LOC approaches, 75
logbooks, 39-41, 189
long-range cruise control, 46-49
Loran-C, 15, 17
low-power cruise control, 49, 51, 52
lowest cost cruise control (LCC), 57-60

M

margin of safety, 163
maximum cruise control, 55-57
mileage, 45-46
minimum en route altitude (MEA), 31
minimum controllable airspeed, 9
multiengine aircraft, 91-93, 169

N

nav mode, 17
navigation, 2
   area navigation (RNAV), 17
   CR mechanical navigation wheel, 34
   inertial, 17
   Loran-C, 17
   OMEGA, 17
   VOR tracking, 15-17
   VOR/DME, 17
NAVSTAR/GPS, 15
NDB approaches, 75, 77-78
new pilots, 102-107
night flight, limitations, aircraft, 91
nonprecision approaches, 74-77

nonstop flight, long-range cruise control, 47
normal cruise control, 52-54
NOTAMs, 18, 123

O

OMEGA, 17
over/out, 134, 135
overspeed propeller, 147
overwater flight
   ditching, 87, 150
      limitations, aircraft, 86-88
   long-range cruise control , 47
   shortwave radio communications, 135

P

pilot, 203-206
pilot-in-command, 205
pitch, 8, 12
pitot heat inoperative, 157
porpoising, 8
precision approaches, (ILS), 13
precautionary shutdowns, 157
precision approach radar (PAR) approaches, 80-81
problem solving, checklist for, 142-144
procedures training, 168
professional aviation, 199
   background for, 200
   commercial flying, 202
   core or basic training for, 201
   differences in training, 200
   instruction for, 202
   pilot vs. copilot, 203-206
   serial learning toward, 200
proficiency, 161-172
   acquisition of, process for, 161-163
   aircraft systems, 167
   checkrides, maintaining proficiency, 163-164
   excuses, egos, rewards for, 171-172
   flight training, 168
   frequency of operation, 165-167
   loss of, 170
   maintaining, 165
   margin of safety, 163
   procedures training, 168
   recurrency training, 164
   regaining, 170-171
   simulator training, 165
propeller, overspeed, 147

R

radar, 2, 15, 28, 80
radio communications, 1
   flight service stations, 17-19
   limitations of, 129
   shortwave, 135
rain, freezing, 194-195
rapid depressurization, 151
rate-of-climb speed, 8, 10
recurrency training, 164
refueling, 38-39
regulations, 2, 22
reserve fuel, 37-38
restricted areas, 2
roger, 133
rough terrain (see inhospitable terrain)
routes, 26-29
rudder, trimming, 3

S

safety, margin of, 163
SDF approaches, 75
separation, aircraft, 2, 4-5, 176
sequence reports, weather, 118
serial learning, 200
shooting approaches, 6, 13
shortwave radio, 135
sidetones and echoes, communications, 131
simulator training, 165
simultaneous transmissions, communications, 140
single-engine aircraft, 90-91, 95, 169
skill or lack thereof, 177, 179
smoke, cabin, 148
snow, 195-197
spins, 154
spirals, 8, 9
squelch, communications, 138
stalls, 22
standard instrument departure (SID), 30
standby, 134
static system clogged, 158
straight-and-level flight, 1, 3
strategy, 183-197
   alternate airports, 190-191
   BILAHs, 185-186
   briefings, 186
   flight logs, 189
   freezing rain, 194-195
   ice, 193-194

IFR, 187-189
snow, 195-197
thunderstorms, 192-193
turbulence, 197
weather, hazardous, 191
synopsis, weather briefings, 116
systems, aircraft, proficiency in, 167

T

takeoff, 7, 10
  alternatives, limitations, aircraft and, 93
  engine out at, 9
  IFR, 95
  urban areas, limitations, aircraft, 89-90
telephone weather briefings, 125-126
temperature aloft, 122, 123
terminal control area (TCA), 2, 28
terrain (see inhospitable terrain)
thunderstorms, 192-193
time of arrival, flight plans, 33-35
traffic advisories, 2
trim, 3, 4, 11
trim runaway, 149
turboprops, limitations, aircraft, 96-97
turbulence, 13, 32, 176, 197
turns, 1, 6-8

U

unable, 134

V

vacuum pump failure, 158
vectored approaches, 75-76
verification of communications, 135
very-low-power cruise control, 49, 51
VFR, 2
  flight plans, 26
  flight recommendations, weather briefings, 116
  limitations, pilot, 103-104
Victor airways, 2
visibility, 35, 107-108
volume, communications, 138
VOR tracking, 2, 6, 15-17, 75
VOR/DME, 17, 65

W

wake turbulence, 176
warning lights, 158
weather, 111-126
  alternate airports, 124-125
  briefings, 114-115
  charts, 111
  FAA standard briefing, 115-124
  forecasting, 113
  freezing rain, 194-195
  hazardous, strategy and, 191-197
  ice, 193-194
  predictions, theory vs. practice, 111-112
  sequence reports, 118
  snow, 195-197
  surface analysis chart, 117
  temperature aloft, 122, 123
  thunderstorms, 192-193
  time element, 113
  turbulence, 197
  weather depiction chart, 119
  winds aloft, 122, 123
weather briefings, 1, 18, 114-124
  adverse conditions, 116
  ATC delays, 124
  computerized, 126
  current conditions, 116
  destination forecast, 120
  en route forecast, 118
  flight service stations and, 17-19
  information-on-request, 124
  limitations, aircraft and, 98
  NOTAMs, 123
  sequence reports, 118
  strategy and, 186
  synopsis, 116
  telephone, 125-126
  VFR flight recommendations, 116
  winds aloft, 121
weather depiction chart, 119
weight and balance, 21-22, 41-43
wilco, 134
winds aloft, 39
  altitude selection and, 32
  time and fuel estimations, 34
  weather briefings, 55, 121, 122, 123